Biomass as Raw Material for the Production of Biofuels and Chemicals

Biomass as Raw Material for the Production of Biofuels and Chemicals

Edited by

Waldemar Wójcik

Faculty of Electrical Engineering and Computer Science,
Lublin University of Technology, Lublin, Poland

Małgorzata Pawłowska

Faculty of Environmental Engineering,
Lublin University of Technology, Lublin, Poland

Routledge
Taylor & Francis Group
LONDON AND NEW YORK

Routledge is an imprint of the Taylor & Francis Group, an informa business

Typeset by codeMantra

Library of Congress Cataloging-in-Publication Data
Names: Wójcik, Waldemar, editor. | Pawłowska, Małgorzata, 1969- editor.
Title: Biomass as raw material for the production of biofuels and chemicals / edited by Waldemar Wójcik, Małgorzata Pawłowska.
Description: Leiden, The Netherlands : Routledge, Taylor & Francis Group, [2022] | Includes bibliographical references.
Identifiers: LCCN 2021031137 | ISBN 9781032011585 (hardback) | ISBN 9781032064574 (paperback) | ISBN 9781003177593 (ebook)
Subjects: LCSH: Biomass energy. | Biomass chemicals. | Biomass.
Classification: LCC TP339 . B5475 2022 | DDC 662/.88—dc23
LC record available at https://lccn.loc.gov/2021031137

Published by: Routledge
Schipholweg 107C, 2316 XC Leiden, The Netherlands
e-mail: Pub.NL@taylorandfrancis.com
www.routledge.com – www.taylorandfrancis.com

ISBN: 9781032011585 (hbk)
ISBN: 9781032064574 (pbk)
ISBN: 9781003177593 (ebk)

DOI: 10.1201/9781003177593

Contents

Preface

Plant biomass, a common source of valuable raw materials, has been used by humans as food, fodder for farm animals, fuel, building and furniture material, as well as a natural medicine or fertilizer for centuries. With the development of civilization, accompanied by the emergence of more efficient energy sources, new structural materials, fertilizers and other chemicals used in various spheres of life, its importance has still not diminished. It is still the basic food for humans and animals, a popular energy source currently used not only as a solid fuel but also – after appropriate processing – as a liquid or gaseous biofuel used in means of transport, a valuable material employed in various industries, as well as a source of bioactive chemicals for the production of pharmaceuticals, nutraceutics, cosmetics or natural agents that improve soil quality.

Today, in addition to the undeniable application values of biomass, special attention is paid to the key role that biomass plays for the Earth's ecosystem, emphasizing its renewable nature, which ensures the circulation of carbon in the global cycle. The growth of biomass is related to the absorption of carbon from the atmosphere *via* photosynthesis. Naturally, the combustion of biomass releases carbon in the form of CO_2, but it can be assumed that the pool of this element in the atmosphere does not increase because it is built up back into the plant tissues. Although treating biomass as a carbon-neutral fuel is an exaggeration, as fossil fuels are also used during the biofuels production, it should be noted that the energetic use of biomass, especially the waste biomass or the mass of hydrobionts such as cyanobacteria, which pose the threat for water ecosystems, certainly contributes to the reduction of the pollutant emissions and provides many other environmental benefits. Such kinds of biomass are especially valuable as a raw material used in biorefineries. The idea of biorefining is gaining more and more popularity around the world. It is based on multidirectional processing of biomass, as a result of which various products are obtained, while maintaining the lowest possible CO_2 emission rate. Biorefining is closely related to another global mainstream concept – the circular economy, in which attention is paid to the fact that by-products generated at various stages of raw material processing are used as substrates in another production process.

Biomass, as a raw material for industry and energy, has a number of advantages including wide availability, renewable nature, and usually low acquisition cost (especially in the case of waste biomass). However, it also has certain disadvantages. Its biodegradable nature can be a problem during transport and storage. Additionally, the use of special preservation methods, such as drying and ensiling, or protection against external factors is sometimes required. On the other hand, in some types of applications, it is necessary to increase the biodegradability of biomass. The high share of polysaccharides and lignin in lignocellulosic structure limits the efficiency of biomass conversion to the targeted products when the biological processing is realized. Enhancement of biodegradability is achieved through a number of processes based on various mechanisms, ranging from the simple mechanical processing consisting in grinding or crushing to complex and multi-stage chemical or physicochemical methods.

The book shows the exemplary applications of different types of biomass for the production of biofuels and other useful products, such as fertilizers, chemicals, and drugs. Special attention is paid to the practical directions of using the biomass of hydrobionts and microorganisms of activated sludge. Considering different applications of the biomass-derived products, the environmental, economic and energetic aspects were taken into account.

Editors

Waldemar Wójcik was born in Poland in 1949. He is the Director of the Institute of Electronics and Information Technology, former long-time dean of the Faculty of Electrical Engineering and Computer Science at Lublin University of Technology, and Doctor Honoris Causa of five universities in Ukraine and Kazakhstan. He obtained his Ph.D. in 1985 at the Lublin University of Technology, and D.Sc. in 2002 at the National University Lviv Polytechnic, Ukraine. In 2009, he obtained the title of professor granted by the President of Poland. In his research, he mainly deals with process control, optoelectronics, digital data analysis and also heat processes or solid-state physics. He pays particular attention to the use of optoelectronic technology in the monitoring and diagnostics of thermal processes. He is a member of Optoelectronics Section of the Committee of Electronics and Telecommunications of the Polish Academy of Sciences and Metrology Section of the Committee of Metrology and Scientific Equipment of the Polish Academy of Sciences. He is also a member of European Academy of Science and Arts (Austria); Academy of Applied Radioelectronics of Russia, Ukraine and Belarus; the International Informatization Academy of Kazakhstan; and many other scientific organizations of Poland as well as Europe and Asia. In total, he has published 56 books and over 400 papers, and authored several patents. He is also a member of the editorial board of numerous international and national scientific and technical journals.

Małgorzata Pawłowska, Ph.D., is a researcher and lecturer at the Faculty of Environmental Engineering of Lublin University of Technology. In 2013–2019, she was the Head of the Department of Alternative Fuels Engineering at the Institute of Renewable Energy Sources Engineering. Currently, she heads the Department of Biomass and Waste Conversion into Biofuels. She received her M.Sc. in philosophy of nature and protection of the environment at the Catholic University of Lublin in 1993. In 1999, she received her Ph.D. in Agrophysics at the Institute of Agrophysics of the Polish Academy of Sciences, and in 2010, she obtained a postdoctoral degree in the technical sciences in the field of environmental engineering at the Wrocław University of Technology. In 2018, she was awarded the title of Professor of Technical Sciences. Her scientific interests focus mainly on the issues related to the reduction of the concentrations of greenhouse gases in the atmosphere, energy recovery of organic waste, and the possibility of using the waste from the energy sector in the reclamation of degraded land. A measurable outcomes of her research is the authorship or co-authorship of 105 papers, including 40 articles in scientific journals, 4 monographs, 24 chapters in monographs, co-edition of 5 monographs, co-authorship of 15 patents and dozens of patent applications. She has participated in the implementation of nine research projects concerning, first of all, the prevention of pollutant emissions from landfills and the implementation of sustainable waste management.

List of Contributors

Roman B. Akselrod
Department of Academic Affairs and Regional Development
Kyiv National University of Construction and Architecture
Kyiv, Ukraine

Yedilkhan Amirgaliyev
Institute of Information and Computational Technologies CS MES RK
Almaty, Kazakhstan

Zhazira Amirgaliyeva
Institute of Information and Computational Technologies CS MES RK
Almaty, Kazakhstan
Faculty of Information Technology
Al-Farabi Kazakh National University
Almaty, Kazakhstan

Mukaddas Arshidinova
Faculty of Information Technology
Al-Farabi Kazakh National University
Almaty, Kazakhstan

Imanbek Baglan
Faculty of Information Technology
Al-Farabi Kazakh National University
Almaty, Kazakhstan

O. V. Bereziuk
Vinnytsia National Technical University
Vinnytsia, Ukraine

Yuriy S. Biks
Faculty of Construction, Thermal Power and Gas Supply
Vinnytsia National Technical University
Vinnytsia, Ukraine

Nataliia O. Bilichenko
Department of Computer Engineering
Vinnytsia National Technical University
Vinnytsia, Ukraine

Victor Bilichenko
Department of Automobiles and Transport Management
Vinnytsia National Technical University
Vinnytsia, Ukraine

Almagul Bizhanova
IT and Control Department
Kazakh Academy of Transport & Communication
Almaty, Kazakhstan

Volodymyr V. Bogachuk
Scientific and Research Department
Vinnytsia National Technical University
Vinnytsia, Ukraine

Gauhar Borankulova
Faculty of Information Technology, Automation and Telecommunications
M.Kh.Dulaty Taraz Regional University
Taraz, Kazakhstan

Anastasiia A. Cherepakha
Department of Life Safety and Safety Pedagogy
Vinnytsia National Technical University
Vinnytsia, Ukraine

Dmytro V. Cherepakha
Department of Construction, Municipal Economy and Architecture
Vinnytsia National Technical University
Vinnytsia, Ukraine

Dmitry I. Denesyak
Green Cool LLC
Vinnytsia, Ukraine

Sergii V. Digtiar
Department of Biotechnology and Bioengineering
Kremenchuk Mykhailo Ostrohradskyi National University
Kremenchuk, Ukraine

Igor N. Dudar
Department of Construction, Municipal Economy and Architecture
Vinnytsia National Technical University
Vinnytsia, Ukraine

Róża Dzierżak
Department of Electronics and Information Technologies
Lublin University of Technology
Lublin, Poland

Yuriy M. Furman
Faculty of Mathematics, Physics, Computer Science and Technology
Vinnytsia Mikhailo Kotsiubynskyi State Pedagogical University
Vinnytsia, Ukraine

Żaklin Grądz
Department of Electronics and Information Technologies
Lublin University of Technology
Lublin, Poland

Konrad Gromaszek
Department of Electronics and Information Technologies
Lublin University of Technology
Lublin, Poland

Kseniya O. Ischenko
Faculty of Civil Engineering, Thermal Power Engineering and Gas Supply
Vinnytsia National Technical University
Vinnytsia, Ukraine

Aigul Iskakova
Institute of Cybernetics and Information Technology
Satbayev Kazakh National Technical University
Almaty, Kazakhstan

Yaroslav V. Ivanchuk
Computer Science Department
Vinnytsia National Technical University
Vinnytsia, Ukraine

Assel Jaxylykova
Faculty of Information technology
Al-Farabi Kazakh National University
Almaty, Kazakhstan
Institute of Information and Computational Technologies CS MES RK
Almaty, Kazakhstan

Aidana Kalabayeva
IT and Control Department
Kazakh Academy of Transport & Communication
Almaty, Kazakhstan

Maksat Kalimoldayev
Institute of Information and Computational Technologies CS MES RK
Almaty, Kazakhstan

Saltanat Kalimoldayeva
Regional Diagnostics Center
Almaty, Kazakhstan

Aliya Kalizhanova
Institute of Information and Computational Technologies CS MES RK
Almaty, Kazakhstan
IT Engineering Department
Kazakhstan University of Power Engineering and Telecommunications
Almaty, Kazakhstan

Gayni Karnakova
Faculty of Information Technology, Automation and Telecommunications
M. Kh. Dulaty Taraz Regional University after
Taraz, Kazakhstan

Zhanna M. Khomenko
Department of Biomedical Engineering and Telecommunications
State University "Zhytomyr Politechnika"
Zhytomyr, Ukraine

Viktoriia O. Khrutba
Department of Ecology
National Transport University
Kiev, Ukraine

Piotr Kisała
Department of Electronics and Information Technologies
Lublin University of Technology
Lublin, Poland

Paweł Komada
Department of Electronics and Information Technologies
Lublin University of Technology
Lublin, Poland

Andrzej Kotyra
Department of Electronics and Information Technologies
Lublin University of Technology
Lublin, Poland

Leonid G. Koval
Biomedical Engineering Department
Vinnytsia National Technical University
Vinnytsia, Ukraine

Nataliia E. Kovshun
Department of Business Economics
National University of Water and Environmental Engineering
Rivne, Ukraine

Ainur Kozbakova
Institute of Information and Computational Technologies CS MES RK
Almaty, Kazakhstan
IT Engineering Department
Almaty University of Power Engineering and Telecommunications
Almaty, Kazakhstan

Mergul Kozhamberdiyeva
Faculty of Information Technology
Al-Farabi Kazakh National University
Almaty, Kazakhstan

Tetyana F. Kozlovs'ka
Kremenchuk Flight College
Kharkiv National University of Internal Affairs
Kremenchuk, Ukraine

Natalia M. Kravets
Institute of Environmental Safety and Monitoring
Vinnytsia National Technical University
Vinnytsia, Ukraine

Liudmyla Kryshtopa
Department of Motor Vehicle Transport
Ivano-Frankivsk National Technical University of Oil and Gas
Ivano-Frankivsk, Ukraine

Sviatoslav Kryshtopa
Department of Motor Vehicle Transport
Ivano-Frankivsk National Technical University of Oil and Gas
Ivano-Frankivsk, Ukraine

Olena Kulakova
Satbayev Kazakh National Technical University
Almaty, Kazakhstan

Serhii M. Kvaterniuk
Institute of Environmental Safety and Monitoring
Vinnytsia National Technical University
Vinnytsia, Ukraine

M. S. Lemeshev
Vinnytsia National Technical University
Vinnytsia, Ukraine

Natalia V. Lyakhovchenko
Faculty of Computer Systems and Automation
Vinnytsia National Technical University
Vinnytsia, Ukraine

Olena G. Lyalyuk
Faculty of Construction, Thermal Power and Gas Supply
Vinnytsia National Technical University
Vinnytsia, Ukraine

Myroslav S. Malovanyy
Department of Ecology and Nature Management
Lviv Polytechnic National University
Lviv, Ukraine

Orken Mamyrbaev
Institute of Information and Computational Technologies CS MES RK
Almaty, Kazakhstan

Oksana V. Maznytska
Department of Biotechnology and Bioengineering
Kremenchuk Mykhailo Ostrohradskyi National University
Kremenchuk, Ukraine

Katharina Meixner
Institute for Environmental Technology
University of Natural Resources and Life Science
Vienna, Austria

Serzhan Mirzabayev
IT and Control Department
Academy of Logistics and Transport
Almaty, Kazakhstan

Vadim P. Miskov
Industrial Engineering Dept.
Vinnytsia National Technical University
Vinnytsia, Ukraine

Mikhailo M. Mushtruk
Faculty of Food Technology and Quality Management of Products of Agriculture
National University Life and Environmental Sciences of Ukraine
Kyiv, Ukraine

Kuanysh Muslimov
Institute of Cybernetics and Information Technology
Satbayev Kazakh National Technical University
Almaty, Kazakhstan

Tatyana N. Nikitchuk
Department of Biomedical Engineering and Telecommunications
Zhytomyr Polytechnic State University
Zhytomyr, Ukraine

Olga V. Novokhatko
Department of Biotechnology and Bioengineering
Kremenchuk Mykhailo Ostrohradskyi National University
Kremenchuk, Ukraine

Dina Nuradilova
Department of Information and Communication Technologies
Asfendiyarov Kazakh National Medical University
Almaty, Kazakhstan

Karlygash Nurseitova
Department of Information and Communication Technologies, Telecommunications
East Kazakhstan State Technical University named after D. Serikbayev
Ust-Kamenogorsk, Kazakhstan

Volodymyr V. Nykyforov
Department of Biotechnology and Bioengineering
Kremenchuk Mykhailo Ostrohradskyi National University
Kremenchuk, Ukraine

Larysa E. Nykyforova
Department of Automation and Robotic Systems named acad. I.I. Martynenko
National University of Life and Environmental Sciences of Ukraine
Kyiv, Ukraine

Olena O. Nykyforova
Department of Biotechnology and Bioengineering
Kremenchuk Mykhailo Ostrohradskyi National University
Kremenchuk, Ukraine

Zbigniew Omiotek
Department of Electronics and Information Technologies
Lublin University of Technology
Lublin, Poland

Ayaulym Oralbekova
Department of Automation, Information Systems and Electric Power Industry in Transport
Kazakh University Ways of Communications
Almaty, Kazakhstan

Sandugash Orazalieva
Institute of Space Engineering and Telecommunications
Almaty University of Power Engineering and Telecommunications (AUPET)
Almaty, Kazakhstan

Ainur Ormanbekova
Faculty of Information Technology
Al-Farabi Kazakh National University
Almaty, Kazakhstan

Igor P. Palamarchuk
Faculty of Food Technology and Quality Management of Products of Agriculture
National University Life and Environmental Sciences of Ukraine
Kyiv, Ukraine

Myroslav Panchuk
Department of Motor Vehicle Transport
Ivano-Frankivsk National Technical University of Oil and Gas
Ivano-Frankivsk, Ukraine

Alona V. Pasenko
Department of Biotechnology and Bioengineering
Kremenchuk Mykhailo Ostrohradskyi National University
Kremenchuk, Ukraine

Marina V. Petchenko
Kremenchuk Flight College
Kharkiv National University of Internal Affairs
Kremenchuk, Ukraine

Roman V. Petruk
Institute of Environmental Safety and Monitoring
Vinnytsia National Technical University
Vinnytsia, Ukraine

Leonid K. Polishchuk
Industrial Engineering Dept.
Vinnytsia National Technical University
Vinnitsa, Ukraine

Saule Rakhmetullina
Department of Information and Communication Technologies, Telecommunications
East Kazakhstan State Technical University named after D. Serikbayev
Ust-Kamenogorsk, Kazakhstan

Georgiy S. Ratushnyak
Faculty of Construction, Thermal Power and Gas Supply
Vinnytsia National Technical University
Vinnytsia, Ukraine

Olga G. Ratushnyak
Faculty of Construction, Thermal Power and Gas Supply
Vinnytsia National Technical University
Vinnytsia, Ukraine

Nataliya V. Rezydent
Faculty of Civil Engineering, Thermal Power Engineering and Gas Supply
Vinnytsia National Technical University
Vinnytsia, Ukraine

Tatyana M. Rotai
Department of Biotechnology and Bioengineering
Kremenchuk Mykhailo Ostrohradskyi National University
Kremenchuk, Ukraine

Oksana A. Sakun
Department of Biotechnology and Bioengineering
Kremenchuk Mykhailo Ostrohradskyi National University
Kremenchuk, Ukraine

Dmitrii M. Salamatin
Department of Biotechnology and Bioengineering
Kremenchuk Mykhailo Ostrohradskyi National University
Kremenchuk, Ukraine

Nataliia B. Savina
Institute of Economics and Management
National University of Water and Environmental Engineering
Rivne, Ukraine

Sergey A. Shargorodsky
Department of Machinery and Equipment of Agricultural Production
Vinnytsia National Agrarian University
Vinnytsia, Ukraine

Indira Shedreyeva
Faculty of Information Technology, Automation and Telecommunications
M. Kh. Dulaty Taraz Regional University
Taraz, Kazakhstan

Valeria S. Shendryk
Department of Biotechnology and Bioengineering
Kremenchuk Mykhailo Ostrohradskyi National University
Kremenchuk, Ukraine

Aigul Shortanbayeva
Faculty of Information technology
Al-Farabi Kazakh National University
Almaty, Kazakhstan

Iryna V. Shvarts
Department of Entrepreneurship, Logistics and Management
Vinnytsia National Technical University
Vinnytsia, Ukraine

Saule Smailova
Department of Information and Communication Technologies, Telecommunications
East Kazakhstan State Technical University named after D. Serikbayev
Ust-Kamenogorsk, Kazakhstan

Andrzej Smolarz
Department of Electronics and Information Technologies
Lublin University of Technology
Lublin, Poland

Christina M. Soloviy
Department of Ecology and Nature Management
Lviv Polytechnic National University
Lviv, Ukraine

Oksana V. Spasichenko
National Transport University
Kiev, Ukraine

Anatoliy I. Svjatenko
Department of Biotechnology and Bioengineering
Kremenchuk Mykhailo Ostrohradskyi National University
Kremenchuk, Ukraine

Aliya Tergeusizova
Faculty of Information Technology
Al-Farabi Kazakh National University
Almaty, Kazakhstan

Stanislav Y. Tkachenko
Faculty of Civil Engineering, Thermal Power Engineering and Gas Supply
Vinnytsia National Technical University
Vinnytsia, Ukraine

Aigul Tungatarova
Kazakhstan, Faculty of Information Technology, Automation and Telecommunications
M.Kh.Dulaty Taraz Regional University
Taraz, Kazakhstan

Ivan S. Tymchuk
Department of Ecology and Nature Management
Lviv Polytechnic National University
Lviv, Ukraine

Oksana A. Ushakova
Technical College
National University of Water and Environmental Engineering
Rivne, Ukraine

Olga Ussatova
Faculty of Information Technology
Al-Farabi Kazakh National University
Almaty, Kazakhstan

Nataliia R. Veselovska
Department of Machinery and Equipment of Agricultural Production
Vinnytsia National Agrarian University
Vinnytsia, Ukraine

Alla P. Vinnichuk
Faculty of Mathematics, Physics, Computer Science and Technology
Vinnytsia Mykhailo Kotsiubynskyi State Pedagogical University
Vinnytsia, Ukraine
Kyiv National University of Construction and Architecture
Kyiv, Ukraine

Waldemar Wójcik
Department of Electronics and Information Technologies
Lublin University of Technology
Lublin, Poland

Olha V. Yavorovska
Department of Construction, Municipal Economy and Architecture
Vinnytsia National Technical University
Vinnytsia, Ukraine

Mykhailo O. Yelizarov
Department of Biotechnology and Bioengineering
Kremenchuk Mykhailo Ostrohradskyi National University
Kremenchuk, Ukraine
Vinnytsia National Technical University
Vinnytsia, Ukraine

Bakhyt Yeraliyeva
Faculty of Information Technology, Automation and Telecommunications
M. Kh. Dulaty Taraz Regional University
Taraz, Kazakhstan

Gulzada Yerkeldessova
Department of Automation, Information Systems and Electric Power Industry in Transport
Kazakh University Ways of Communications
Almaty, Kazakhstan

Mykhaylo V. Zagirnyak
Department of Electromechanics
Kremenchuk Mykhailo Ostrohradskyi National University
Kremenchuk, Ukraine

Volodymyr V. Zhurav
Department of Biotechnology and Bioengineering
Vinnytsia National Technical University
Vinnytsia, Ukraine

Sergii M. Zlepko
Biomedical Engineering Department
Vinnytsia National Technical University
Vinnytsia, Ukraine

Vadym I. Zyuzyun
Department of Ecology
National Transport University
Kiev, Ukraine

Chapter 1

The Intensity of Heat Exchange in Complexes of Organic Waste Disposal

Stanislav Y. Tkachenko, Kseniya O. Ischenko, Nataliya V. Rezydent, and Leonid G. Koval
Vinnytsia National Technical University

Dmitry I. Denesyak
Green Cool LLC

Roman B. Akselrod
Kyiv National University of Construction and Architecture

Konrad Gromaszek
Lublin University of Technology

Serzhan Mirzabayev
Academy of Logistics and Transport

Aigul Tungatarova
M. Kh. Dulaty Taraz Regional University

CONTENTS

1.1 INTRODUCTION

The increase in energy efficiency of biogas plants is slowed down by the drawbacks of the methods, structures and technologies for thermal stabilization of the mixture in the bioreactor, which causes problems with temperature constancy throughout the reactor volume. It is well known that biogas plant (BGP) in the west produces more energy in the form of biogas than is required to maintain the functioning of the BGP itself (pre-heating, transportation, heat stabilization, etc.). Another problem concerns the raw materials for BGP, as it can be animal, agricultural, food and

DOI: 10.1201/9781003177593-1

industrial wastes, and usually even mixtures of all of these. The number of variants of mixture compositions is infinite; so the study of each variant is not appropriate. This was confirmed in the conditions announced for scientific works at the XVI Minsk International Forum on Heat and Mass Exchange (http://www.itmo.by/conferences/mif), which clearly indicates that the research of the thermophysical properties of substances is not accepted and has no scientific value. For the qualitative course of the fermentation process, a rather strict compliance with the temperature mode at BGP is required. In this case, under the conditions of acceptable temperature fluctuations in the psychrophilic mode amount to ±2°C, and in thermophilic mode, the accuracy increases to ±0.5°C (Sadchikov and Kokarev, 2016). An important role is played by indirect parameters such as fraction size, substrate moisture in different seasons, and mixing intensity. Devising a rational mixing method contributes to the creation of optimal hydrodynamic and temperature conditions for the existence of the methane-forming bacteria, as well as a more efficient use of a digestion tank volume (Tropin, 2011). This leads to the development of means for evaluating the thermophysical properties and intensity of heat exchange of raw materials at the BGP, directly during the process of fermentation. There are no such plants and experiments that would cover the full range of problems described above. In practice, the methods for assessing the intensity of heat exchange under the conditions of limited information on the thermophysical properties of complex mixtures are required in order to implement both deep fundamental research and rapid analysis at the existing BGP. Such methods have emerged and developed (Tkachenko and Pishenina, 2017), and the convergent results obtained with these methods are available. It is now possible to further improve these techniques using the regular thermal mode (RTM) (Kondratev, 1954; Osipova, 1979).

The purpose of this chapter is to show how to increase the energy efficiency of heat exchange equipment by reducing the uncertainty of estimating the intensity of heat exchange in liquid multiphase mixtures, which are prone to structural changes, by improving the methods and means of implementing the experimental calculation method using the RTM. To achieve this objective, the following tasks were set: analysis of experimental data obtained on a test rig; development of a mathematical model to evaluate the effect of structure cracking and mixing on the intensification of the heat transfer process; development of recommendations to use this technique to estimate heat transfer intensities during the development of full-scale heat exchangers.

1.2 MATERIAL AND RESEARCH RESULTS

The solid fraction of pig manure was selected as a test substance. The approximate weight of a fattening pig is 85kg. According to NTP 17-99kh (2001), under the described conditions, the humidity accounts for 75%. For research, solid manure was stirred with water to obtain different humidity. The stirring process was implemented in the following sequence: using the balance dependencies, the required mass of water to be added to a certain mass of solid fraction was calculated, then the determined amounts of manure and water were mixed in the experimental tank, with thorough stirring of the solution until homogeneous concentration was obtained. The following experiments were enabled with the finished mixture – substrate.

Substrate is considered to be a non-homogeneous mixture, the dispersion medium of which are water-soluble salts and low-molecular-mass organic compounds of animal excretion, while the dispersed phase includes solids and insoluble impurities of mineral and organic origins.

The substrates with humidity of 82%, 85%, 90%, and 94% were used in the studies. The measurements were conducted after 5 and 20 days after the substrates were prepared. The animals were fed with peas with bran. The content of dry organic substances for this manure was not calculated.

During the preparation of the substrates, precipitation occurred in the solution with a moisture content of 82%, and poor homogeneity was observed, even with constant stirring. In a mixture of 85%, some heterogeneity was discovered, and the precipitate was observed 5 minutes after stopping stirring. For the mixtures with a moisture content of 90% and 94%, homogeneity distribution of components with the formation of a blurry viscous slurry and no precipitation about 15 minutes after stopping stirring were observed. A mixture of 85%–94% humidity was selected for research at the experimental setup.

The studies were performed using the experimental setup described in Tkachenko and Pishenina (2017). It consisted of an inner tank having the form of a thin-walled cylinder (wall thickness 0.5 mm), in which the mixture was heated. Heating occurred due to the heat of the water located in the outer coaxial channel. The annular cavity had an outer diameter of 200 mm, the inner one was 97 mm, the height was 12 cm and the inner cylinder had a diameter of 96 mm and height of 90 mm.

The ambient temperatures were measured by the height of the water and substrate tanks at five points in the geometric center. The weight of hot water was 2.3 kg, and the weight of the test liquid amounted to 700–800 g. The average heating time was 20 minutes. The total amount of heating free convection conditions was 20, and with forced conditions, it was 40.

The scheme of the experimental setup is displayed in Figure 1.1. The characteristics of the test environment are shown in Table 1.1.

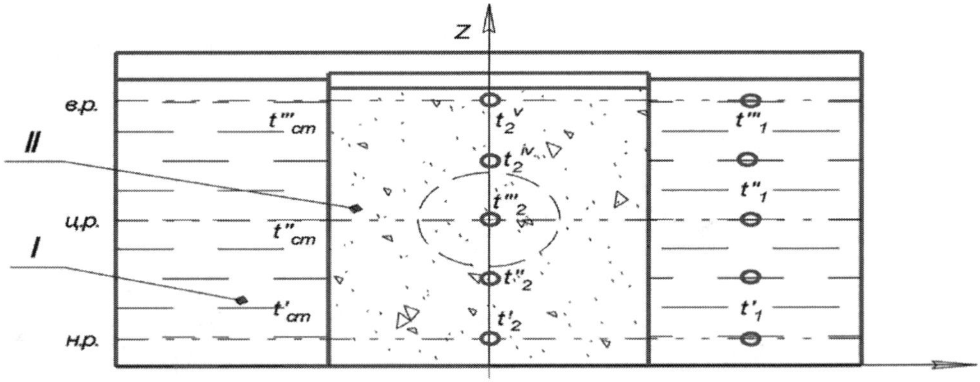

Figure 1.1 Schematic diagram of temperature measurements t_1 and t_2.

Table 1.1 Characteristics of the test environments

Line number	*Liquid environment*	*Type of convection*	*Heating tempo*	*Shear rate, 1/s*	*The range of change of average volume water temperature*	*The range of change of medium volume temperature*	*Thermophysics properties of the test environment*
(1)	*(2)*	*(3)*	*(4)*	*(5)*	*(6)*	*(7)*	*(8)*
1	Water	Free	0.0068		82 ... 25	22 ... 60	Known
2	Sunflower oil	Forced	0.0017	0.53	82 ... 25	22 ... 60	Known
3	Sugar solution, 70%	Forced	0.0016	0.53	75 ... 30	30 ... 58	Known
4	Pig manure substr., W = 94%	Free	0.00064	–	75 ... 30	25 ... 60	Unknown
5	Pig manure substr., W = 90%	Free	0.0005	5–	75 ... 30	25 ... 60	Unknown
6	Pig manure substr., W = 85%	Free	0.00051	–7	5 ... 30	25 ... 60	Unknown
7	Pig manure substr., W=90%	Forced	0.0013	1.07	75 ... 30	25 ... 60	Unknown
8	Pig manure substr., W = 90%	Forced	0.001	0.85	75 ... 30	25 ... 60	Unknown
9	Pig manure substr., W = 90%	Forced	0.0008	0.53	75 ... 30	25 ... 60	Unknown
10	Pig manure substr., W = 94%	Forced	0.0036	1.07	75 ... 30	5 ... 60	Unknown
11	Pig manure substr., W = 94%	Forced	0.0027	0.85	75 ... 30	25 ... 60	Unknown
12	Pig manure substr., W = 94%	Forced	0.0016	0.53	75 ... 30	5 ... 60	Unknown

According to Figure 1.1, I is the external cavity with heating environment (water), II is the inner cylindrical tank with test environment, $t_1^{I} \ldots t_1^{III}$ are local temperatures in the outer water tank, $t_2^{I} \cdots t_2^{V}$ are local temperatures in the inner tank with the test environment, and $t_{st}^{I} \ldots t_{st}^{III}$ was obtained from the wall temperature calculation.

The sequence of the series of experiments was as follows: filling the internal cavity of the experimental unit (Figure 1.1) with the test fluid (manure, sugar solution, oil, and water) and filling the outer cavity with hot water. During the experiment, the water in the outer tank heated the liquid in the inner tank.

For research, an experimental calculation method was employed, improved by applying the theory of RTM (Tkachenko and Denesiak, 2017). This technique using RTM is convenient in this case because the thermal stabilization of the plant is not needed for its implementation, it is capable of measuring temperatures at two or more points, it is transient and as a consequence of lower requirements for thermal isolation, there is a wide choice of heat sources. The required values are the coefficients of heat exchange from the water to the wall α_1 and from the wall to the environment α_2, and the heating rate of the environment m.

The distribution of excessive temperature over time for humidity of the manure substrate amounting to 92% and 95% is shown in an example (Figure 1.1). Excessive temperature is the modulus of difference between the average volume of liquid environment $t_{p.c}$ in the inner vessel (I) and the average volume of water in the annular channels (II) t_B:

$$\vartheta = \left| \overline{t_{p.c}} + \overline{t_B} \right| \tag{1.1}$$

where $\overline{t_{p.c}} = \left(t_1^{I} + t_1^{II} + t_1^{III} + t_1^{IV} + t_1^{V} \right) / 5$; $\overline{t_B} = \left(t_2^{I} + t_2^{II} + t_2^{III} + t_2^{IV} + t_2^{V} \right) / 5$°C.

In order to determine average volume temperatures, averaging of the temperatures along the height of the heat exchange surface was implemented at fixed times (Figure 1.2).

The concept of heating or cooling to explain the obtained data is used. In the literature, (Kondratev, 1954; Kukharchuk et al., 2016; Osipova, 1979), it is stated that an RTM occurs when function 2 acquires a linear appearance:

$$m = \frac{\ln \vartheta - \ln \vartheta}{\tau' - \tau''}, \tag{1.2}$$

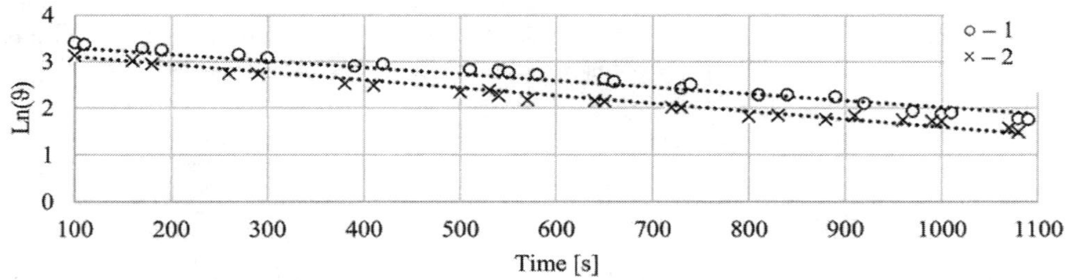

Figure 1.2 Excessive temperature 1 – pig manure with a humidity of 92%; 2 – pig manure with a humidity of 94%.

where

ϑ' and ϑ'' are the excessive local body temperatures at the initial τ' and the final τ'' time, consequently, $\vartheta = t_1'' - t_{ct}$, where t_1'', t_{ct} are determined for two time points τ' and τ'. At the RTM, the substrate with a humidity content of 94% has $m = 0.0017$, while for humidity 90%$-m = 0.0015$. The issues are obtained as a result of statistical processing of the experimental data. The data approximation for substrates with a different humidity gives a determination coefficient of 97%–98%.

Figure 1.3, for comparison, shows the experimental results for water, sugar, and sunflower oil. The processing was implemented similarly to the previous one. The studies were conducted in the same temperature range and on the same experimental stand. The temperature measurement results were displayed on a computer (Kukharchuk et al., 2016; Kukharchuk et al., 2017; Osadchuk et al., 2018).

Let us present, for a more complete analysis of the research materials, the dependence of heating rate of the mixture on its parameters (humidity, concentration) (Table 1.1).

Table 1.1 makes it possible to analyze the information as following: a substrate with a humidity content of up to 90% is weakly intensified by heat exchange with stirring, $m = 0.0013$. In turn, a 2% increase in humidity gives an increase in the heating rate which is two to three times more substantial than stirring. This means that an increase in solids suppresses the heat exchange to a greater extent than convection.

The data allow rapid assessment of the propensity of different mixtures to heat exchange, which is one of the stages of analysis of the existing plants operation for the plugging of organic waste. Using the indicator m – heating rate – it is possible to determine whether it is appropriate to use such a mixture and implement its heating, as well as to pre-analyze and establish what is appropriate to add to the mixture to increase or decrease its ability to heating.

An important issue in the quality of BGP work is stirring. The choice of the optimum velocity range can be estimated by considering the rheological behavior of the fluid with limited information on thermophysical properties. The experimental values of the complex of physical properties of the liquid $(\text{ECPP}_6)^{(\text{Vn})}$ obtained at the basic experimental setup (Tkachenko et al., 2018) were used to evaluate the rheological structure of the test liquids.

In order to estimate the increase in the heat exchange intensity under the conditions of the destruction of substrate structure, the technique proposed in Tkachenko et al. (2018) was employed, through a criterion equation describing the heat exchange intensity in the internal working cavity. Equation 1.3 for forced motion with natural convection is obtained by our experimental setup described in Tkachenko and Pishenina (2017):

$$Nu = 0.0549 \cdot \text{Re}^{*0.589} \cdot \text{Pr}_p^{0.33} \cdot \left(\text{Gr}_H \cdot \text{Pr}_p\right)^{0.1} \cdot \left(\frac{\text{Pr}_p}{\text{Pr}_{st}}\right)^{0.25} \tag{1.3}$$

In equation 1.3: $\text{Re}^* = \overline{w} \cdot 2\delta / \nu$ – Reynolds number for our conditions; Pr_p – Prandtl number for the liquid, determined by the liquid temperature; $\text{Gr}_H = \left(g\beta\overline{\Delta t}H^3\right)/\nu^2$ is the Grashoff number; Pr_c denotes the Prandtl number for liquid, determined by the wall temperature; $\overline{w} = \pi \cdot n \cdot d_M / 60$ – conditional characteristic velocity of the liquid

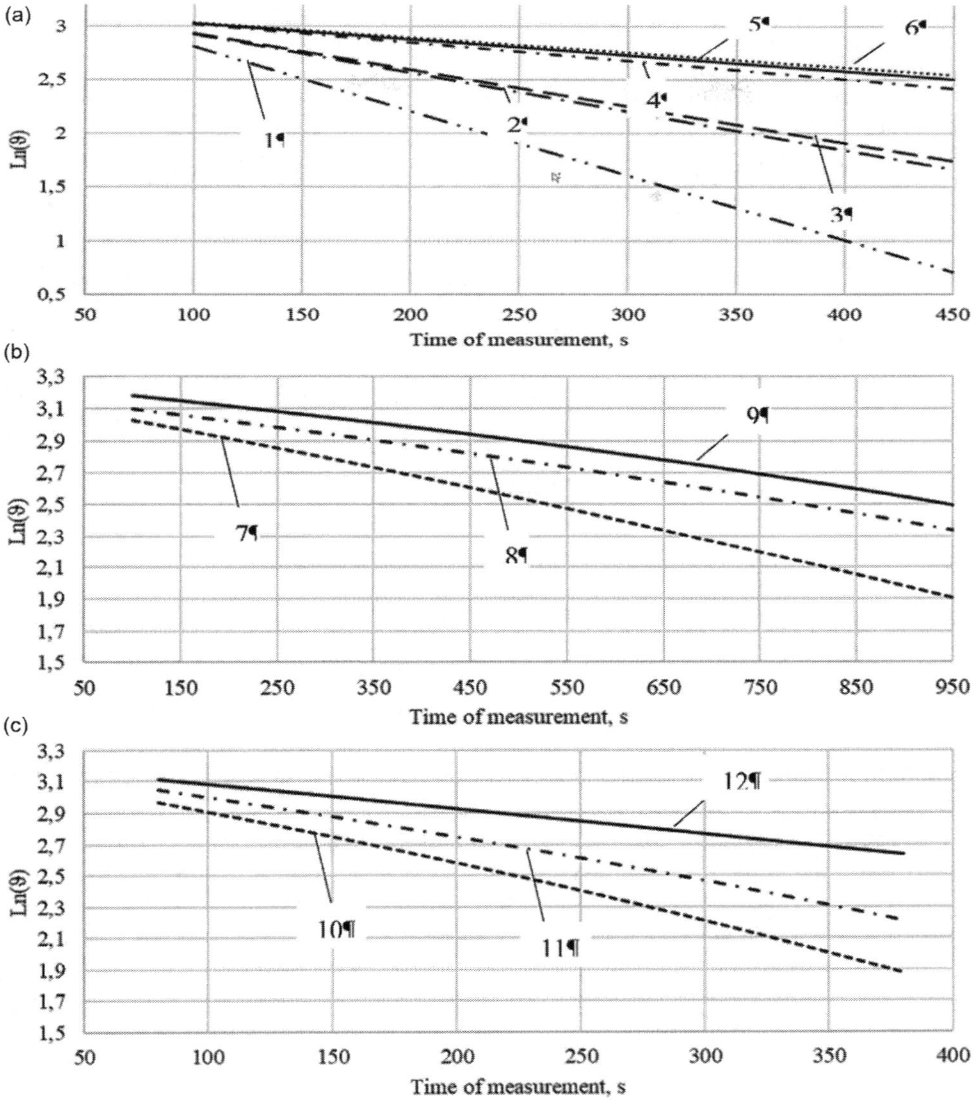

Figure 1.3 Excessive logarithms of excessive temperatures of the test liquid environments. a – heterogeneous liquids; b – pig manure substrate, $W = 90\%$; C – pig manure substrate, $W = 94\%$ 1 – water; 2 – sunflower oil; 3 – sugar solution, mass concentration of 70%; 4 – pig manure substrate, $W = 94\%$, free convection; 5 – pig pus substrate, $W = 90\%$, free convection; 6 – pig manure substrate, $W = 85\%$; 7–9 – pig manure substrate, $W = 90\%$, shear rate 1.07, 0.85, 0.53 (1/s), respectively; 4 – pig pus substrate, $W = 94\%$, shear rate 1.07, 0.85, 0.53 1/s (1/s), consequently.

motion, m/s; n is the speed of the stirrer, rpm; d_M is determining the linear size for the forced convection, the diameter of the stirrer m; ν is the kinematic viscosity of the liquid environment, m²/s; g is the free fall acceleration, m/s²; H is the determining linear size in natural convection conditions, m; $\overline{\Delta t} = \left(\overline{t_{st}} - \overline{t_p}\right)$ is the temperature head; $\overline{t_{st}}$ and $\overline{t_p}$ are the average temperature of the wall and the liquid environment under study, respectively, °C; β is the coefficient of temperature expansion of the liquid environment, °C^{-1}; λ is the thermal conductivity of the liquid environment, W/(m K); ρ is the density of the liquid environment, kg/m³; c_p – specific heat of liquid environment, kJ/(kg K). The correction for the direction of heat exchange $\left(\Pr_p/\Pr_{st}\right)_\sigma^{0.25}$ was determined by a specially developed method under the conditions of use of liquids and mixtures whose thermophysical properties are unknown (Tkachenko et al., 2018).

Equation 1.3 can be used within $20 < \mathrm{Re}^*_{2\delta} < 3.7103$, $6.106 < (\mathrm{Gr}_H \cdot \Pr_p) < 2.108$, $3.2 < \Pr_p < 1.7103$.

From equation 1.3, we determine the basic complex of physical properties (CPP) in the following form:

$$\mathrm{CPP}_6 = C_p^{0.43} \cdot \rho^{0.43} \cdot \beta^{0.1} \cdot \lambda^{0.57} \cdot \nu^{-0.359} \tag{1.4}$$

$$P_1 = C_p^{0.43} \cdot \rho^{0.43} \cdot \beta^{0.1} \cdot \lambda^{0.57} \tag{1.5}$$

In order to estimate the growth of heat exchange intensity during the destruction of the structure from the array of experimental data obtained during processing, separate the results under the conditions $\overline{t_2} = \text{const}$, $\overline{\Delta t_\sigma} = \text{const}$, and represent in the form of dependence (Figure 1.4). The shear rate equivalent for the conditions of a particular plant $\gamma = \overline{w}/\left[0.5\left(D_{BH} - d_M\right)\right]$ is taken according to Tkachenko et al. (2018), and α_2 is the coefficient of heat exchange from the heat exchange surface to the liquid environment under study, W/(m² K).

In Figure 1.4, curves 3 and 4 represent the approximation of the experimental data for the humidity of 90% and 94% as a function $\alpha_2 = C \cdot \gamma^b$, where b is the approximation exponent and C is the coefficient of equation. The obtained curves are linear, since points for construction are chosen in conditions $\overline{t_2} \approx \text{const}$ and $\overline{\Delta t} \approx \text{const}$.

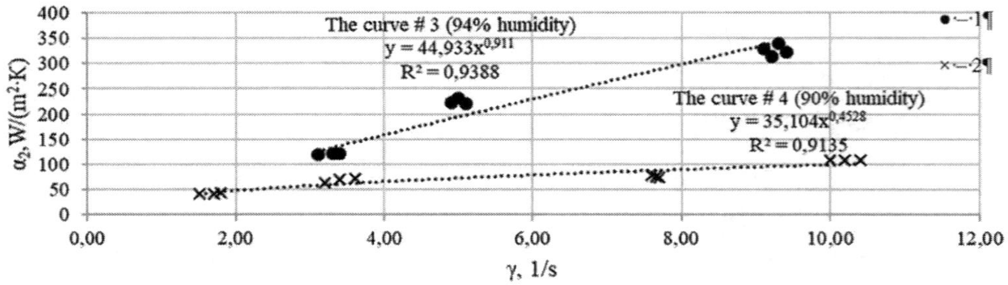

Figure 1.4 The dependence of the heat exchange coefficient on the substrate on the shear rate: 1 – pig manure with humidity 94%; 2 – pig manure with humidity 90%; $t_2 = 42$; $\Delta t = 11°\text{C}–13°\text{C}$; 3, 4– degree approximation of the test data.

For pig manure with 90% humidity, the same analysis shows the increase in the heat exchange coefficient by 1.7 times, but for a 90% humidity substrate, the initial dependence for α_2 does not correspond to our experimental data. The structure destruction is observed in this case, but the destructive effect is similar as under the condition $W=90\%$ calculated taking into account the dependence (3) and does not show such a high proportion of the increase in the heat exchange intensity under the condition of the structure destruction, as in $W=94\%$. While implementing data analysis with the choice of others $\overline{t_2} = \text{const}$, $\overline{\Delta t_\sigma} = \text{const}$ shows similar results within the expected error of 15%–20%.

To experimentally determine the coefficient of heat transfer from the wall to the investigated liquid medium, the experimental value of the complex of physical properties from equations 1.3–1.5 is as follows:

$$\text{ECCP}_6 = \frac{\alpha_2^{\text{eksp}}}{0.0549 \cdot \underbrace{\left[\overline{w}^{-0.59} \cdot \left(g \cdot \overline{\Delta t_\sigma}\right)^{0.1} \cdot \frac{H^{0.3}}{2\delta^{0.41}}\right]}_{P_{\sigma 2}} \cdot \left(\frac{\text{Pr}_p}{\text{Pr}_{ct}}\right)_\sigma^{0.25}} \tag{1.6}$$

The analysis conducted by means of the numerical methods of dependences (1.3) and (1.6) in the studied range of temperature changes and $\dot{\gamma}$ showed that the degree of change of the complex parameters P_1 is significantly lesser than the degree of change $v^{-0.36}$, which is consistent with Osipova (1979). The experimental data are presented in schedule $\left(\text{ECCP}_6\right)^{(1/n)} / \left(\text{ECCP}_6\right)_{\max}^{(1/n)} = f\left(\bar{\dot{\gamma}}\right)$ (Figure 1.5), the equivalent of effective viscosity (Tkachenko et al., 2018).

The effect of the structure destruction and the stirring speed was evaluated with the dependence:

$$\alpha' / \alpha_0 = \left(\gamma' / \gamma_0\right)^{n1} \cdot \left(\Delta t / \Delta t_0\right)^{0.25}, \tag{1.7}$$

where α' and α_0 are the current and minimum values of the heat exchange coefficient to the substrate, W/(m^2 K); γ' and γ_0 are the current and minimum values of the shear rate equivalent; $\Delta t'$ and Δt_0 – current and minimum temperature heads; $\left(\Delta t / \Delta t_0\right)^{0.25} \approx \left(\text{Pr}_p / \text{Pr}_{ct}\right)^{0.25}$ is introduced to evaluate the influence of the heat exchange direction.

Dependence (1.7) follows dependence (1.3), if we take into account that the above-mentioned dependencies relate to the experimental results obtained at the same plant at the same determining temperature within the slight variations of the temperature head between the heating wall and the liquid environment.

Figure 1.5a shows the tendency described in the previous article, the general tendency of changing the complex of physical properties with decreasing the shear rate – gamma.

In our opinion, temperature head in the experiment, the different chemical and granulometric composition, structure, its state, the ratio of liquid, and solid and gaseous phases influenced the different behaviors of pig manure compared with the previous work.

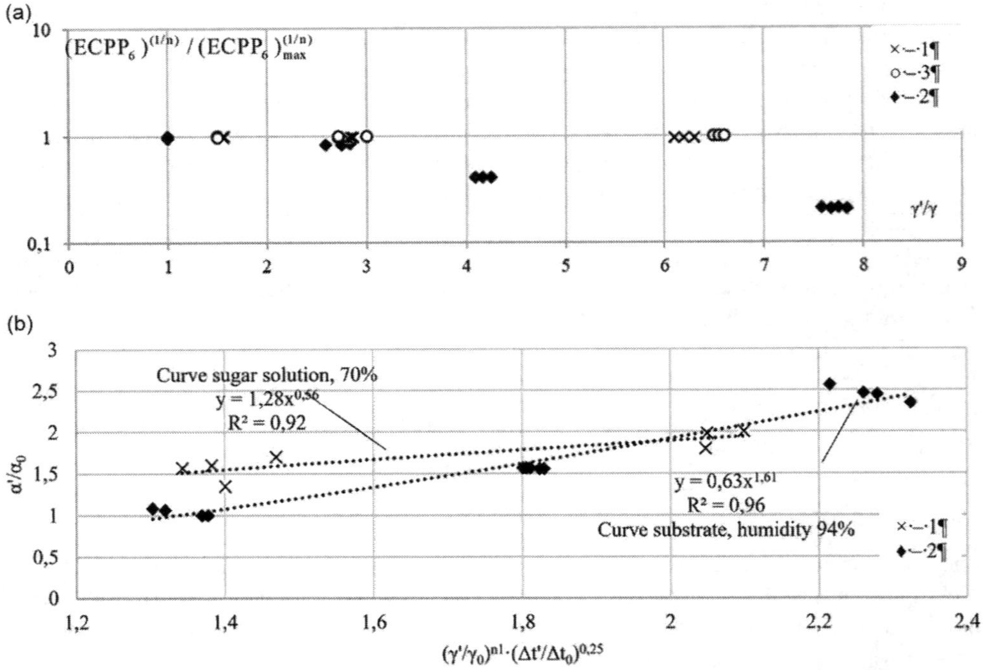

Figure 1.5 Caption experimental results of the effective viscosity equivalent under different values of the condensed shear rate of the test liquid: 1 – sugar solution with a mass concentration of 70%, $\overline{\Delta t_\sigma}$ = 7°C–18°C; 2 – pig manure with humidity of 94%, temperature head between hot and cold coolant in the basic experiment $\overline{\Delta t_\sigma}$ =7–18; 3 – water, $\overline{\Delta t_\sigma}$ = 5°C–20°C. The average water temperature, sugar solution, and pig manure is ~40°C. (a) The effective viscosity equivalent. (b) Evaluation of the effect of structure destruction and stirring speed on the heat exchange intensity.

While implementing an analysis of the dependence $\alpha'/\alpha_0 = (\gamma'/\gamma_0)^{n1} \cdot (\Delta t/\Delta t_0)^{0.25}$, it was found that the closest liquid in terms of heat exchange properties to this substrate is a 70% mass concentration of sugar solution. As can be seen from the figure, for a given sugar solution, the exponent in the equation is 0.56, which corresponds to the exponent in equation 1.6. Since there is a structure destruction in the substrate with a humidity content of 94%, the exponent near γ'/γ_0 is 1.61, which indicates the feasibility of intensifying the heat exchange of the substrate by stirring.

The data obtained are of value in the calculation of BGP and the choice of stirring system, thermodynamic processes in biotechnology.

1.3 CONCLUSIONS

The increase in the energy efficiency of biogas plants is slowed down by the disadvantages of methods, structures, and technologies for thermal stabilization of the mixture in the bioreactor, and the problems of temperature constancy over the reactor volume.

The studies of heat exchange were conducted in structured and unstructured liquid media with limited information on physical properties, under the conditions of the non-stationary mode of heat exchange. Data processing was carried out simultaneously as stationary and non-stationary heat exchange processes (method of regular heat mode) which allowed us to establish the feasibility of using the method and the conditions under which the liquid mixture is destroyed.

It was established that the manure is a structured fluid with 90% and 94% humidity. The experimental conditions under which the breakage of the structure occurs and, as a result, a significant increase occurs in the heat exchange intensity, were established.

REFERENCES

Kondratev, G.M. 1954. *Regulyarny'j teplovoj rezhim*. Moskva: Gosudarstvennoe izdatelstvo tekhni-ko-teoreticheskoj literatury.

Kukharchuk, V.V., Hraniak, V.F., Vedmitskyi, Y.G., Bogachuk, V.V., Zyska, T., Komada, P., Sadikova, G. 2016. Noncontact method of temperature measurement based on the phenomenon of the luminophor temperature decreasing; *Proceedings of the SPIE*, 28 September 2016, 10031–100312F. Photonics Applications in Astronomy, Communications, Industry, and High-Energy Physics Experiments 2016, Wilga, Poland.

Kukharchuk, V.V., Kazyv, S.S., Bykovsky, S.A. 2017. Discrete wavelet transformation in spectral analysis of vibration processes at hydropower units. *Przeglad Elektrotechniczny*, 93(5): 65–68.

Osadchuk, A., Osadchuk, V., Baraban, S., Zyska, T., Zhanpeisova, A. 2018. Temperature transducer based on metal-pyroelectric-semiconductor structure with negative differential resistance; *Proceedings of the SPIE 10808, 108085D*, 1 October 2018. Photonics Applications in Astronomy, Communications, Industry, and High-Energy Physics Experiments 2018, Wilga, Poland.

Osipova, V.A., 1979. *Eksperimentalnoe issledovanie proczessov teploobmena. Ucheb. posobiedlyavuzov*. Moskva: Energiya.

Sadchikov, A., Kokarev, N. 2016. Optimizacziya teplovogo rezhima v biogazovy'kh usta-novkakh; *Fundamental 'ny' eissledovaniya*, 2-1: 90–93.

Tkachenko, S., Denesiak, D. 2017. Perspektyvy vykorystannia metodiv rehuliarnoho rezhymu dlia vyznachennia intensyvnosti teploobminu v obmezhenomu obiemi. *Suchasni tekhnolohii materialy I konstruktsii v budivnytstvi. Nauk.-tekhn. zbirnyk*, 2(23): 106–112.

Tkachenko, S., Pishenina, N.V. 2017. *Novi metody vyznachennia intensyvnosti teploobminu v systemakh pererobky orhanichnykh vidkhodiv*. Vinnytsia: VNTU.

Tkachenko, S., Palamarchuk, N., Denesiak D. 2018. Teplofizychne testuvannia reolohichnoho povodzhennia skladnykh ridynnykh seredovyshch. *Visnyk Vinnytskoho politekhnichnoho instytutu*, 4: 46–53.

NTP 17-99kh. 2001. *Normy tekhnologicheskogo proektirovaniya system udaleniya I podgotovki k ispolzovaniyu navoza I pometa*. Moskva: Minsel`khozprod.

Tropin, A.N. 2011. Povyshenie effektivnosti raboty samotechnoj sistemy udaleniya navoza putem optimizaczii ee konstruktivnykh i tekhnologicheskikh parametrov. Dissertacziya kandidata tekhnicheskikh nauk, Sankt-Peterburg-Pavlovsk, 31: 50–63.

Chapter 2

Predicting Volume and Composition of Municipal Solid Waste Based on ANN and ANFIS Methods and Correlation-Regression Analysis

Igor N. Dudar, Olha V. Yavorovska, and Sergii M. Zlepko
Vinnytsia National Technical University

Alla P. Vinnichuk
Vinnytsia Mykhailo Kotsiubynskyi State Pedagogical University
Kyiv National University of Construction and Architecture

Piotr Kisała
Lublin University of Technology

Aigul Shortanbayeva
Al-Farabi Kazakh National University

Gauhar Borankulova
M. Kh. Dulaty Taraz Regional University

CONTENTS

DOI: 10.1201/9781003177593-2

2.1 INTRODUCTION

Currently, the issues of forecasting the amount of municipal solid waste (MSW) generation and the analysis of its morphological composition constitute an important task for creating an effective environmental waste management system in the cities. The objective data on MSW volumes and composition are critical for a well-founded policy of solid waste management. Obtaining reliable information about the volume of MSW generation in the settlements should ensure effective planning and control over the waste management system. On the other hand, the analysis of the MSW morphological composition will enable prompt changes to the scheme of primary collection, sorting, transportation, and what is more important – it will help choose the method for processing resource-valuable waste and organic fractions, as well as the final method for the residual MSW disposal.

One of the important characteristics of the MSW generated in settlements includes not just MSW volume but also its morphological composition, which represents the ratio of individual components: cardboard and paper, glass, metal, plastic, construction waste, and some other fractions that are part of MSW. The statistical information on the volume of resource-valuable fractions in the total flow of MSW generated by citizens will make it possible to forecast the estimated level of recycling at both local and regional levels. This, in turn, will allow for efficient MSW processing, which will become a criterion for saving resources and will enable to develop goals of the circular- or closed-loop economy. The concept of the circular economy presupposes saving as much costs of resources, products, and materials as it is possible in order to create the products with long service life and, thus, increasing the sustainability of the world and city economy as well as contributing to the implementation of Paris Agreement and the UN Sustainable Development Goals. Therefore, it is the process of MSW processing, as one of the elements of waste management system that is an important step in establishing and entrenching the principles of the closed-loop economy.

2.2 ANALYSIS OF LITERARY SOURCES AND PROBLEM STATEMENT

The dependence of processing efficiency on the accuracy of forecasting primary MSW collection was noticed more than 20 years ago (Chang and Lin 1997). We have thoroughly analyzed the existing methods. All literature methods for predicting the volume of MSW generation can be divided into several groups:

- Methods based on time series (Denafas et al. 2014, Mwenda et al. 2014, Bridgewater 1986);
- Methods based on deterministic factor and stochastic correlations (Shan 2010, Kolekar et al. 2016, Kumar and Samadder 2017);
- Methods of GIS cluster analysis (Owusu-Sekyere et al. 2013, Purcell and Magette. 2009, Thanh et al. 2010);
- Methods of statistical learning theory (Bandara et al. 2007, Chen and Chang 2000, Vu 2015).

See Table 2.1 for the advantages and disadvantages of the existing methods.

Today, the time-series methods are the most common. They involve the assessment of the waste fluctuation trends over time. These methods are relevant because they

Table 2.1 Characteristics of the municipal solid waste forecasting methods

Method	*Advantages*	*Disadvantages*
Methods based on time series	• Low labor costs of forecasting due to a simple procedure of baseline data collection; • Illustrative nature of results; • Efficiency in case of short-term planning	• Inaccurate results due to impact factors being neglected; • Methods cannot be used for long-term planning; • The need for multiple observations under the same conditions; • Methods cannot be used to assess the dynamics of morphological composition; • Lack of empirical justification of the forecast trend; • When using these methods, it is not possible to take into account the future trends
Methods based on deterministic factor and stochastic correlations	• The ability to identify hypothetical casual relationships between impact factors; • Subjective nature of the study, which allows identifying the most important factors influencing the overall municipal solid waste flow	• Methods cannot be used to assess the dynamics of morphological composition; • Subjective nature of the study, which requires obtaining expert assessment to confirm the adequacy of the results obtained; • Only demonstrates a causal relationship; • Labor input involved in multiple observations and formation of a single database with informative data; • It is impossible to respond to dramatic changes in the trends of forecasting factors that are likely to occur during the study of the social and behavioral factors; • When using these methods, it is not possible to take into account future trends
Methods of GIS cluster analysis	• Illustrative nature of the obtained research result; • Study results are easy to understand; • More than two variable parameters can be forecast	• It is impossible to demonstrate dynamics over time; • Forecasting is impossible due to the static results obtained; • Lack of a model for forecasting; • When using these methods, it is not possible to take into account the future trends
Methods of statistical learning theory	• It is possible to detect non-linear data dependencies; • More than two variable parameters can be forecast; • The method does not provide for obtaining multiple experiment attempts; • It is not important to select the same conditions, the system responds and perceives deviations from the main trend; • High forecasting accuracy	• The complexity of forecasting due to the need for thorough and long preparation, the presence of a local minimum, and difficulties in determining the network-based architecture

are easy to assess. Despite having numerous drawbacks, this method is often used by scientists and researchers, as well as employees of municipal and public organizations (Bridgewater 1986, Kotsyuba 2013, Mwenda et al. 2014).

Methods based on deterministic factor and stochastic correlations. These methods are aimed not only at forecasting the MSW volumes but also at identifying the hypothetical causal relationships between the factors predicting the MSW generation. These methods are widely used to forecast the level of solid waste generation based on the socio-economic factors and other variables in the studies by Oribe-Garcia et al. (2015), Chung and Lin (1997), Kotsiuba (2013), and Owusu-Sekyere et al. (2013).

GIS cluster analysis methods (Purcell and Magette 2009): An interesting example of solving the problem of forecasting the MSW volume and morphological composition is the use of modern GIS (Purcell and Magette 2009). The advantage of cluster analysis is that it can simultaneously analyze more than two parameter characteristics. However, it is difficult to demonstrate the trend of MSW generation and changes in its morphological composition over time using the cluster analysis. Therefore, the cluster analysis cannot be used to forecast the future trends of each individual characteristic, which also renders it inapplicable as a decision-making tool for optimizing the sphere of waste management system.

Methods of statistical learning theory: All the above-mentioned forecasting methods have one significant drawback: the uncertainty of their results – the model forecasts are never certain. Today, the methods of statistical learning theory are the most accurate in terms of forecasting. These models first run the input-output data, whereafter they are able to identify their relationships, find the dependencies between them (i.e., the learning process), and then forecast the future results (Abbasi et al. 2013). Considering all the advantages and disadvantages of the existing groups of methods for forecasting the MSW volumes and morphological composition, the methods of statistical learning theory are deemed the best.

Recently, many researchers have used the machine learning and artificial intelligence methods in their studies of solid waste management – Noori et al. (2010), Azadi and Kamiri-Jashni (2016).

Today, there are few studies conducted with the application of Artificial Neural Networks (ANN) to the field of solid waste management (Abdoli et al. 2011). For instance, ANN was used to forecast weekly solid waste in the study by Noori et al. (2010). The greatest attention to the use of ANN for forecasting the MSW volumes was given by the scientists from the University of Iran – Azadi and Karimi-Jashni (2016), Shahabi et al. (2012). Thus, Azadi and Karimi-Jashni (2016) applied ANN and multilinear regression models using the temperature, height, population count, and frequency of waste collection as input variables for forecasting seasonal waste generation in Iran. However, to date, the subject of all studies was the mixed waste flow, and there are still no published studies that would be dedicated to forecasting the morphological composition of waste.

2.3 PURPOSE AND TASKS OF RESEARCH

The information above allows us to note that the methods of statistical learning theory are the most effective in forecasting the MSW volumes and morphological composition. Unlike other methods, it ensures high forecasting accuracy and does not involve a great multitude of experiment attempts under the same conditions, which is not possible for the social and behavioral effects on the MSW generation. In addition, these are the only methods that will enable forecasting the morphology, rather than a component-by-component composition of MSW, since this category of methods makes it possible to forecast more than two output parameters.

Moreover, a literature analysis has shown that to date, there are no detailed studies on forecasting the MSW morphological composition using any of the known methods. Therefore, to obtain reliable research results, we have set the following goals:

1. Analyze the factors that affect the volume of MSW generation and its morphological composition.
2. Create a mathematical model for describing the volume of MSW generation and its morphological composition.
3. Establish an effective dependence of the MSW generation on the social and behavioral factors, factors of living conditions, organizational factors, and so on, by comparing the existing methods of deterministic factor correlation and statistical learning theory.

2.4 MATERIALS AND METHODS

The research methodology is shown in Figure 2.1. For the analysis, we collected the daily data on the MSW volume and morphological composition in the households of Vinnytsia (Ukraine). The study was conducted in 2015–2018. In total, over 700 qualitative samples were collected that allowed determining the dependence of the MSW morphological composition on the social behavior of city residents, which was done for the first time. In order to search for an effective forecasting model, we created three mathematical models using the methods of statistical learning theory – ANN, Adaptive Neuro-Fuzzy Inference Systems (ANFIS), and by compiling correlation and regression models. These models were compared with each other in order to select the most effective model by analyzing the measurement errors: the determination coefficient. In order to assess data, we used the Neural Network Toolbox module of the MatLab package.

2.5 SELECTION OF MODEL FACTORS

The morphological composition of waste is not static. It is constantly changing under the influence of natural, socio-economic, or other factors. When it comes to MSW, its morphological composition depends on the social and behavioral situation of the citizens generating it. This change can be predetermined by a variety of reasons: citizen behavior, increased social culture or responsibilities, their habits, change of diet, expanding the family, income, and a number of other factors. In the study, we tried to consider the most typical changes that can happen to households in the city.

The impact factors were selected to further analyze the dependencies between the generation of specific MSW fractions and a number of social and economic factors. A formal and logistic method was used to select these factors. We identified the indirect

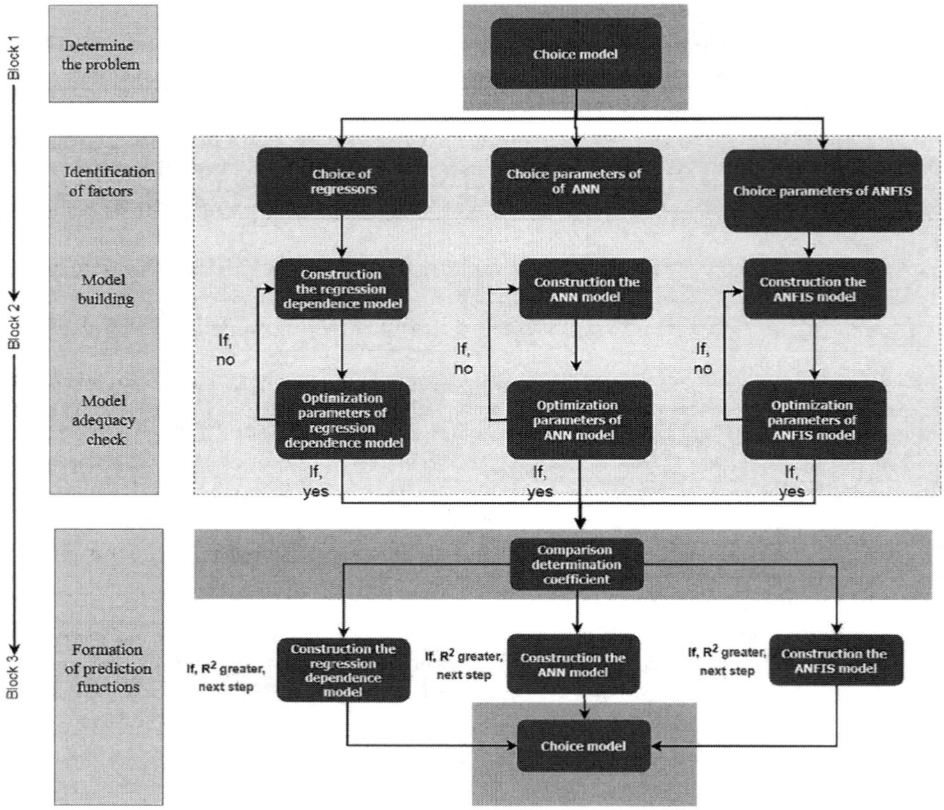

Figure 2.1 Research algorithm.

and direct impact factors. Figure 2.2 features a diagram of the direct and indirect impact factors influencing the changes in the morphological composition of MSW.

Note that none of the factors can be viewed in isolation since their interdependence is constantly growing.

Each of the considered factors can have both a positive and a negative effect, which predetermines the need to analyze them in detail for the forecasting purposes.

At the same time, the social and behavioral factors are among the most difficult to analyze. However, it is important to forecast them, since they will have the greatest impact on the waste morphological structure. There is a phrase "Show me your waste and I will tell you who you are". Thus, the social and behavioral factors may only be analyzed by means of a questionnaire and a survey. It is not possible to obtain adequate data for correlation dependencies using other methods of the scientific experiment. This group of factors has become a decisive factor at the time of research method selection (Bereziuk et al. 2019).

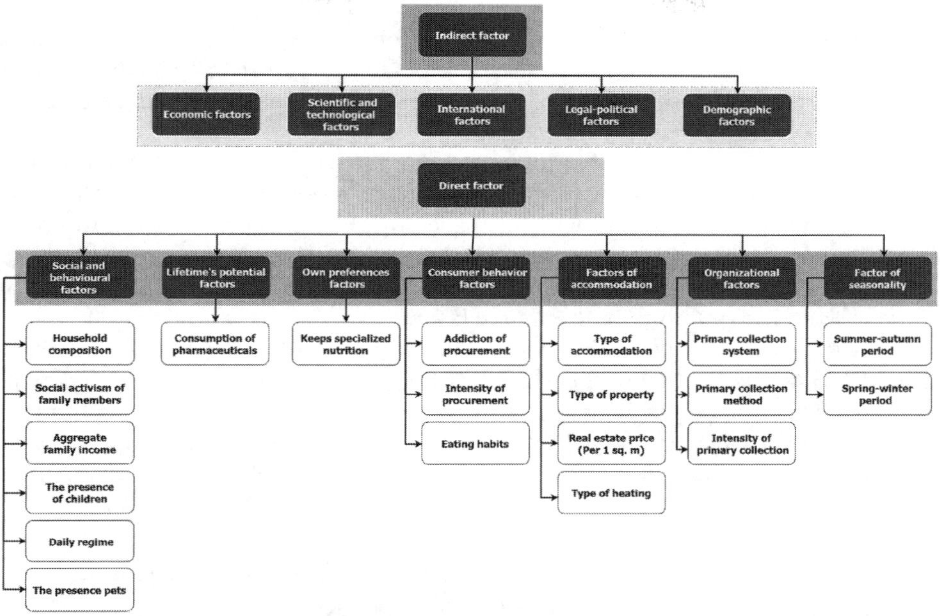

Figure 2.2 Diagram of the direct and indirect impact factors influencing the morphological composition of municipal solid waste.

2.6 CREATION OF A MATHEMATICAL MODEL USING STATISTICAL LEARNING THEORY AND CORRELATION AND REGRESSION METHOD

For forecasting, we selected the methods of ANN, ANFIS, and correlation-regression dependencies.

The efficiency of the ANN and ANFIS methods is assessed and compared in terms of the nonlinear error functions that are statistically significant and measure the distribution of errors. An ANN is a computer model of a "black box" originating from a simplified concept of the human brain, such as the ability to learn, think, remember, and solve problems. The ANN architecture consists of one input layer, hidden layers, and one output layer where each layer consists of simple processing elements called neurons that are connected to other neurons in the next layer and, therefore, form different ANN types. Among the learning algorithms, backpropagation is the most reliable learning algorithm in the neural network that works based on a gradient descent to minimize errors during each iteration. The Levenberg-Marquardt backpropagation algorithm gives the most satisfactory forecasting results.

ANFIS is an adaptive fuzzy inference network authored by Jang. ANFIS is a multilayer backpropagation network where each node performs a specific signal reception function and has a set of parameters related to that node. Like ANN, ANFIS can reflect and forecast unknown input data to their results by learning rules from the already known data.

In this paper, socio-behavioral factors, factors of life potential, self-preference, living conditions, organizational and seasonality factors were defined as the model input data, while the values of resource-valuable MSW fractions and fractions that are not recyclable were defined as the output data.

Figure 2.1 shows the ANN architecture and learning result using ten neurons in a hidden layer. The network was implemented and modeled using the Matlab R2013b software and the nn function, which is based on the Levenberg-Marquard optimization for learning with redistributing. The tan-sigmoid transfer function (tansig) was applied with the reverse layer propagation algorithm on the hidden layer and linear transfer function (purelin) on the source layer.

An ANFIS fuzzy neural network used to forecast the morphological composition of MSW was created in the Matlab R2013b environment using the Fuzzy Logic Toolbox extension package and ANFIS editor. This allowed identifying hidden patterns in the data and forming a database of fuzzy inference rules based on the findings (Figure 2.3). The result of the composition of a correlation-regression model is performed using Excel 2010.

2.7 COMPARISON OF THE ANN AND ANFIS MODELS, AND CORRELATION AND REGRESSION ANALYSIS

In order to select the most effective forecasting model based on ANFIS, ANN, and correlation-regression analysis methods, it is necessary to compare their results with each other. It is best to select a forecasting model based on measurement errors. One of the known factors that can be used in such comparisons of forecasts is the determination coefficient and the mean square error (MSE). The determination coefficient is a statistical indicator that provides certain information on the extent to which the dependable variable variation depends on the independent variable variation. In other words, it is a statistical indicator of how well the model forecasts the actual data points. A higher determination coefficient indicates that the model matches the data better. MSE is a mean square error between actual and forecast data. The R_2 coefficient and MSE for three models are shown in Table 2.2.

On the basis of what was explained earlier, higher values of the determination coefficient would indicate a better capacity of the model in forecasting specific characteristics under study. According to Table 2.2, the R2 value for the ANN model is 0.920, which indicates the capabilities of this model in assessing the weight of individual solid waste fractions. In addition, the R2 value is determined as 0.917 for the ANIFIS model, which indicates that this model is a suitable forecasting model. In general, the R_2 values for both ANN and ANFIS models are high, so both models are able to forecast the volume and morphological composition of MSW. In addition, a comparison of these two models shows that the R_2 value for the ANN model is higher than for the ANFIS model. As a result, the ANN model is more capable of forecasting than ANFIS. Thus, the R_2 error for the proposed ANN model is 0/920.

A comparison of the ANFIS error with the correlation and regression analysis model shows that this value is 46% higher than the best regression model applied to the

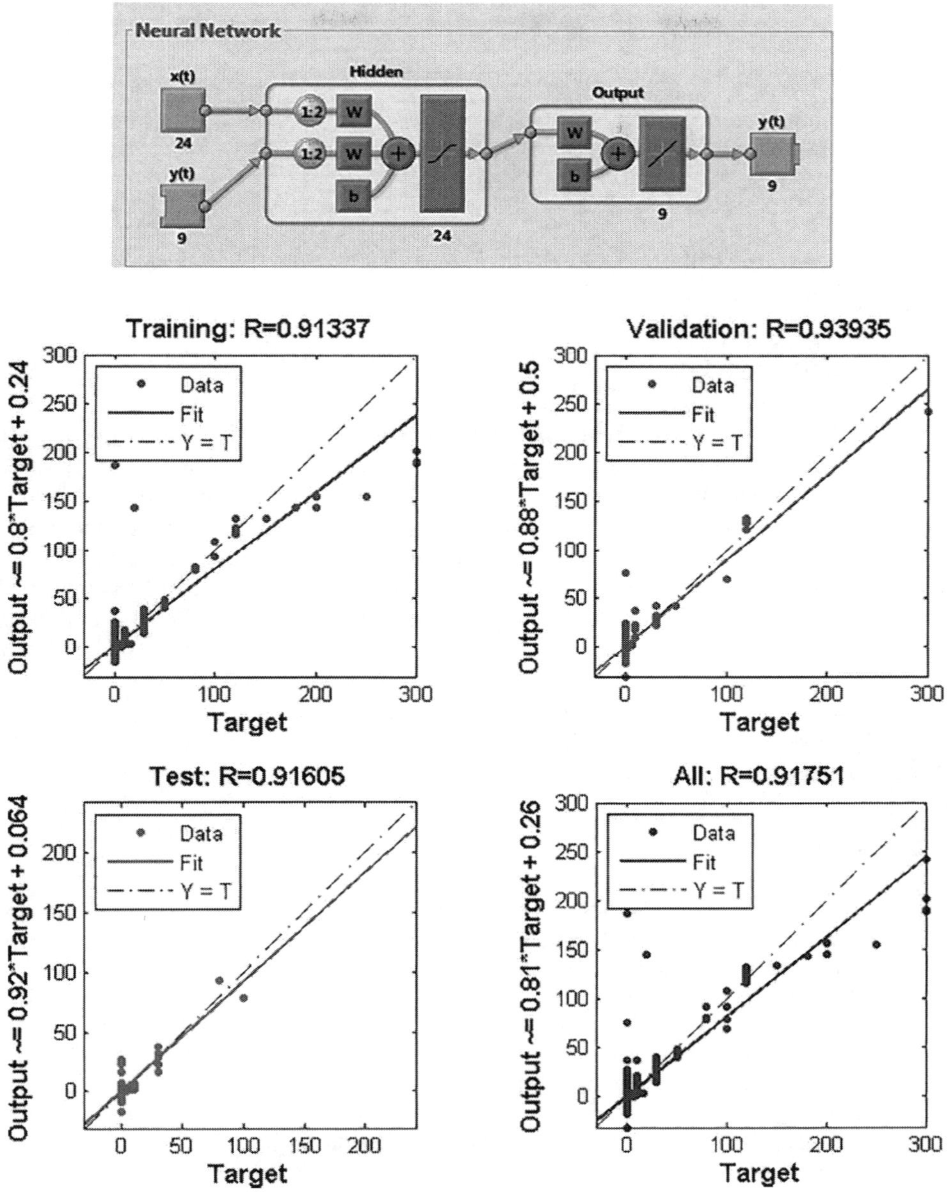

Figure 2.3 ANN architecture and learning result.

same data set ($R_2 = 0.63$), which indicates that ANN has a significantly greater forecasting capacity than the traditional regression methods. Another advantage is that ANN is not as sensitive to non-standard changes as the regression model. However, ANN is a more complex method, and model developing and results interpreting are more

Table 2.2 The results of errors

Model	R_2	*MSE*
Regression	0.920	0.0237
ANN	0.917	0.0223
ANFIS	0.630	0.0131

difficult. However, ANN's higher performance makes it a valuable tool in determining the strategies for increasing recycling and achieving the circular economy goals.

2.8 CONCLUSIONS

Forecasting the rate of waste generation is not an easy task. This is mainly due to a lack of up-to-date data and a rapid change of external factors (inflation, political stability, changes in legislation on manufacturer liability). An even more difficult task is to forecast the morphological composition of MSW, which is influenced by an even greater number of factors, including social and behavioral factors that are difficult to forecast.

In the study, we identified the factors that influence the volume of MSW generation and its morphological composition, namely: social and behavioral factors, factors of life potential, personal preference, living conditions, organizational and seasonal factors.

Forecasting was carried out using ANN and ANFIS and correlation-regression dependencies. The final choice of the forecasting method was made by comparing the model's error. The ANN model is the best in terms of forecasting, which is due to its forecasting accuracy and relatively easy application.

REFERENCES

Abbasi, M., Abduli, M.A., Omidvar, B., Baghvand, A. 2013. Results uncertainty of support vector machine and hybrid of wavelet transform-support vector machine models for solid waste generation forecasting. *Environmental Progress & Sustainable Energy* 33(1): 220–228. DOI:10.1002/ep.11747

Abdoli, A., Falah Nezhad, M., Salehi Sede, M., Behboudian, S.R. 2011. Longterm forecasting of solid waste generation by the artificial neural networks. *Environmental Progress & Sustainable Energy* 31(4): 628–636. DOI:10.1002/ep.10591

Azadi, S., Karimi-Jashni, A. 2016. Verifying the performance of artificial neural network and multiple linear regression in predicting the mean seasonal municipal solid waste generation rate: A case study of Fars province, Iran. *Waste Management* 48, 14–23.

Bandara, N.J.G.J., Hettiaratchi, J.P.A., Wirasinghe, S.C., Pilapiiya, S. 2007. Relation of waste generation and composition to socio-economic factors: A case study. *Environmental Monitoring and Assessment* 135(1–3), 31–39. DOI:10.1007/s10661-007-9705-3

Bereziuk, O. et al. 2019. Ultrasonic microcontroller device for distance measuring between dustcart and container of municipal solid wastes. *Przegląd Elektrotechniczny* 95(4) 146–150. DOI:10.15199/48.2019.04.26

Bridgewater, A.V. 1986. Refuse composition projections and recycling technology. *Resources and Conservation* 12: 159–174.

Chang, N.B., Lin, Y.T. 1997. An analysis of recycling impacts on solid waste generation by time series intervention modeling. *Resources, Conservation and Recycling* 19: 165–186.

Chen H.W., Chang N.B. 2000. Prediction analysis of solid waste generation based on grey fuzzy dynamic modeling. *Resources, Conservation and Recycling* 29: 1–18.

Denafas, G. et al. 2014. Seasonal variation of municipal solid waste generation and composition in four East European cities. *Resources, Conservation and Recycling* 89: 22–30.

Kolekar, K.A., Hazra, T. Chakrabarty, S.N. 2016. A review on prediction of municipal solid waste generation models. *Procedia Environmental Sciences* 35: 238–244.

Kotsyuba, I.G. 2013. Rationale for scientific works on the reliable production and transportation of solid waste (on the example of Zhytomyr): Dissertation, Kyiv.

Kumar, A., Samadder, S.R. 2017. An empirical model for prediction of household solid waste generation rate – A case study of Dhanbad, India. *Waste Management* 68: 3–15.

Mwenda, A., Kuznetsov, D., Mirau, S. 2014. Time series forecasting of solid waste generation in Arusha city – Tanzania. *Mathematical Theory and Modelling* 4(8): 29–39.

Noori, R., Karbassi, A., Salman Sabahi, M. 2010. Evaluation of PCA and Gamma test techniques on ANN operation for weekly solid waste prediction. *Journal of Environmental Management* 91(3): 767–771.

Oribe-Garcia, I. et al. 2015. Identification of influencing municipal characteristics regarding household waste generation and their forecasting ability in Biscay. *Waste Management* 39: 26–34.

Owusu-Sekyere, E., Harris, E., Bonyah, E. 2013. Forecasting and planning for solid waste generation in the Kumasi metropolitan area of Ghana: An ARIMA time series approach. *International Journal of Sciences* 2: 69–83.

Purcell, M., Magette, W.L. 2009. Prediction of household and commercial BMW generation according to socio-economic and other factors for the Dublin region. *Waste Management* 29(4): 1237–1250. DOI:10.1016/j.wasman.2008.10.011

Shahabi, H., Saeed, K., Ahmed, B.B., Zabihi, H. 2012. Application of artificial neural network in prediction of municipal solid waste generation (Case study: Saqqez 157 City in Kurdistan Province). *World Applied Sciences Journal* 20(2): 336–343.

Shan, C.S. 2010. Projecting municipal solid waste: The case of Hong Kong SAR. *Resources, Conservation and Recycling* 54: 759–768.

Thanh, N.P., Matsui, Y., Fujiwara, T. 2010. Household solid waste generation and characteristic in a Mekong Delta city, Vietnam. *Journal of Environmental Management* 91: 2307–2321.

Vu, H. L. 2015. Advanced numerical modeling techniques for modern waste management systems. A thesis submitted to the Faculty of Graduate Studies and Research in Partial Fulfillment of the Requirements for the Degree of Doctor of Philosophy in Environmental Systems Engineering, University of Regina.

Chapter 3

Assessment of Ecology-Economic Efficiency in Providing Thermal Stabilization of Biogas Installations

Georgiy S. Ratushnyak, Olena G. Lyalyuk, Olga G. Ratushnyak, Yuriy S. Biks, and Iryna V. Shvarts
Vinnytsia National Technical University

Roman B. Akselrod
Kyiv National University of Construction and Architecture

Paweł Komada and Żaklin Grądz
Lublin University of Technology

Kuanysh Muslimov
Satbayev University

Olga Ussatova
Al-Farabi Kazakh National University

CONTENTS

3.1 INTRODUCTION

One of the ways of environmental-friendly and rational use of fuel and energy resources is the investment in the renewable energy sources, such as construction of solar, wind, and bioelectric power plants. In Ukraine, such investments were about 3.7 million euros in 2019. The implementation of bioconversion helps dispose of organic waste in biogas installations, prevents the contamination of the biosphere with harmful substances, and allows obtaining an alternative source of energy – biogas. The process of methane formation requires energy consumption for thermal stabilization of anaerobic fermentation in biogas installations (Suresh et al. 2013, Weiland 2003). In terms of scientific research, the energy-saving mechanisms for providing thermal stabilization in biogas energy installations are insufficiently substantiated.

DOI: 10.1201/9781003177593-3

It is possible to increase the energy efficiency and environment friendliness of bioconversion by rational selection of alternative renewable heat sources for fermentation processes of thermal stabilization. Such alternative sources are solar energy, low-potential thermal energy of soil and water as well as utilization of thermal emissions of bioconversion systems (Zabarnyi & Shurchkov 2002). Ukraine plans to give up the coal energy by 2050. The share of renewable energy sources is predicted to make up 70%.

The analysis of the literature shows that the theoretical and experimental studies of animal waste bioconversion mechanisms and kinetics of the technological process were conducted by G. Ratushnyak, E. Larushkin, S. Yakushko, M. Drukovani, and O. Zuev (Rotshtein 1999, Noyola et al. 2006, Rotshtein et al. 2008). An expert system for an intelligent support of energy-saving management of bioconversion technological process is also considered in the works of E. Larushkin and A. Rotshtein (Rotshtein 1999, Rotshtein et al. 2008). There are no examples in the literature of the calculations related to the ecological and economic efficiency of bioconversion (Drukovani et al. 2006, Noyola et al. 2006, Rotshtein et al. 2008). The purpose of the research was to create a theoretical background and develop a scientifically proven system of making effective decisions for choosing an innovative bioconversion project.

In order to achieve this goal, the research should solve the following tasks (Geletukha & Martseniuk 1999, Ratushnyak & Anokhina 2013, Zhelikh et al. 2013):

- Develop a classification of factors affecting ecological and economic indicators of the mechanisms for ensuring thermal stabilization in biogas plants (Zadeh 1975, Zuev 2009).
- Develop a hierarchical system of mathematical models related to multifactor analysis regarding the ecological economic efficiency of the mechanism ensuring the process of thermal stabilization based on fuzzy logic, which takes into account the influence of quantity and quality factors (Redko et al. 2016, Kukharchuk et al. 2017).

3.2 METHODOLOGY

The classification of the factors affecting ecological and economic indicators of mechanisms ensuring thermal stabilization in biogas plants was developed. The inference tree (Figure 3.1) is based on classified factors; it defines a system of nested statements and establishes hierarchical links between them. It characterizes the impact of a set of influencing factors. It can be represented as a fixed ratio:

$$Y = f(X,Z), \tag{3.1}$$

where

X is a linguistic variable which describes environmental factors;

Y is a linguistic variable which describes economic factors.

The linguistic variable X that describes environmental factors can be represented by the expression:

$$X = f\left(x_1, x_2, x_3\right) \tag{3.2}$$

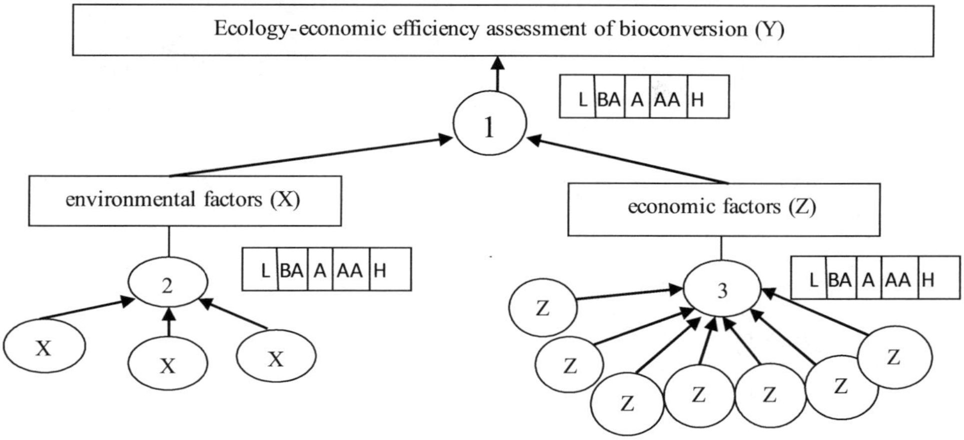

Figure 3.1 The tree of inference which concerns hierarchical relationships of factors that influence environmental and economic assessments of a bioconversion project.

where
x_1 is the linguistic variable "CO_2 emissions",
x_2 is the linguistic variable "organic waste disposal",
x_3 is the linguistic variable "environmental pollution".

The linguistic variable Z, which describes environmental factors, can be represented by the expression:

$$Z = f\left(z_1, z_2, z_3, z_4, z_5, z_6, z_7\right), \tag{3.3}$$

where
z_1 – the linguistic variable "term of recoupment";
z_2 – the linguistic variable "Net Present Value";
z_3 – the linguistic variable "Internal Rate of Return";
z_4 – the linguistic variable "Profitability Index",
z_5 – the linguistic variable "duration of operation";
z_6 – the linguistic variable "profit";
z_7 – the linguistic variable "operating costs".

Simulation of the system-level intellectual support of project variant selection can be done by using the following terms:

$$T(Y) = \langle \text{Low, Below Average, Average, Above Average, High} \rangle$$
$$T(X) = \langle \text{Low, Below Average, Average, Above Average, High} \rangle$$
$$T(Z) = \langle \text{Low, Below Average, Average, Above Average, High} \rangle$$

The fuzzy inference technique helps calculate the predicted number by means of a fuzzy set using the "IF - THAT" linguistic expression system. It combines the fuzzy terms of output and input variables by using operations I and OR, which correspond to operations min and max (Zadeh 1975, Rotshtein et al. 1997, Rotshtein 1999).

Linguistic statements correspond to a system of fuzzy logical equations, which characterize the surface of belonging to the variables (X, Z) of the corresponding term:

$$\mu_L(Y) = \mu_L(X) \wedge \mu_L(Z) \vee \mu_{BA}(X) \wedge \mu_L(Z) \vee \mu_L(X) \wedge \mu_{BA}(Z) \tag{3.4}$$

$$\mu_{BA}(Y) = \mu_{BA}(X) \wedge \mu_{BA}(Z) \vee \mu_{BA}(X) \wedge \mu_A(Z) \vee \mu_A(X) \wedge \mu_{BA}(Z) \tag{3.5}$$

$$\mu_A(Y) = \mu_A(X) \wedge \mu_A(Z) \vee \mu_L(X) \wedge \mu_{AA}(Z) \vee \mu_{AA}(X) \wedge \mu_L(Z) \tag{3.6}$$

$$\mu_{AA}(Y) = \mu_{AA}(X) \wedge \mu_{AA}(Z) \vee \mu_{AA}(X) \wedge \mu_A(Z) \vee \mu_A(X) \wedge \mu_{AA}(Z) \tag{3.7}$$

$$\mu_H(Y) = \mu_H(X) \wedge \mu_H(Z) \vee \mu_{AA}(X) \wedge \mu_H(Z) \vee \mu_H(X) \wedge \mu_{AA}(Z) \tag{3.8}$$

The assessment of linguistic variables levels related to environmental factors (X) is done by the amount of CO_2 emissions (x_1), the level of organic waste utilization (x_2) and the level of environmental pollution (x_3) using the following system of term sets:

$$T(x_1) = \langle \text{Low, Average, High} \rangle$$
$$T(x_2) = \langle \text{Low, Average, High} \rangle$$
$$T(x_3) = \langle \text{Low, Average, High} \rangle$$

The system of fuzzy logical equations which characterizes the surface of belonging to the variables (x_1, x_2, x_3) of the corresponding term can be shown as the linguistic statement:

$$\mu_L(X) = \mu_H(x_1) \wedge \mu_L(x_2) \wedge \mu_H(x_3) \vee \mu_H(x_1) \wedge \mu_L(x_2) \wedge \mu_A(x_3) \vee \mu_A(x_1) \\ \mu_L(x_2) \wedge \mu_H(x_3) \tag{3.9}$$

$$\mu_{BA}(X) = \mu_A(x_1) \wedge \mu_H(x_2) \wedge \mu_H(x_3) \vee \mu_A(x_1) \wedge \mu_L(x_2) \wedge \mu_A(x_3) \vee \mu_H(x_1) \cdot \\ \mu_L(x_2) \wedge \mu_A(x_3) \tag{3.10}$$

$$\mu_A(X) = \mu_A(x_1) \wedge \mu_A(x_2) \wedge \mu_A(x_3) \vee \mu_H(x_1) \wedge \mu_L(x_2) \wedge \mu_A(x_3) \vee \mu_L(x_1) \cdot \\ \mu_H(x_2) \wedge \mu_A(x_3) \tag{3.11}$$

$$\mu_{AA}(X) = \mu_A(x_1) \wedge \mu_A(x_2) \wedge \mu_L(x_3) \vee \mu_A(x_1) \wedge \mu_H(x_2) \wedge \mu_A(x_3) \vee \mu_L(x_1) \cdot \\ \mu_H(x_2) \wedge \mu_A(x_3) \tag{3.12}$$

$$\mu_H(X) = \mu_L(x_1) \wedge \mu_H(x_2) \wedge \mu_L(x_3) \vee \mu_L(x_1) \wedge \mu_H(x_2) \wedge \mu_L(x_3) \vee \mu_A(x_1) \\ \mu_H(x_2) \wedge \mu_L(x_3) \tag{3.13}$$

The assessment of the levels of linguistic variables linking economic factors (Z) with payback period (z_1), with Net Present Value (z_2), with Internal Rate of Return (z_3), with Profitability Index (z_4), with duration (z_5), with profit (z_6), and with operating costs (z_7) is performed by using a system of term sets:

$$T(z_1) = \langle \text{Small(S), Average, Long(Ln)} \rangle$$
$$T(z_2) = \langle \text{Low, Average, High} \rangle$$
$$T(z_3) = \langle \text{Low, Average, High} \rangle$$
$$T(z_4) = \langle \text{Low, Average, High} \rangle$$
$$T(z_5) = \langle \text{Small(S), Average, Long(Ln)} \rangle$$
$$T(z_6) = \langle \text{Low, Average, High} \rangle$$
$$T(z_7) = \langle \text{Low, Average, High} \rangle$$

The system of fuzzy logical equations which characterizes the surface of belonging to the variables (z_1, z_2, z_3, z_4, z_5, z_6, z_7) of the corresponding term can be shown as the linguistic statement:

$$\begin{aligned} &\mu_L(Z) = \wedge \mu_{Ln}(z_1) \wedge \mu_L(z_2) \wedge \mu_s(z_3) \wedge \mu_L(z_4) \wedge \mu_L(z_5) \wedge \mu_L(z_6) \wedge \mu_L(z_7) \vee \\ &\mu_A(Z_1) \wedge \mu_L(z_2) \wedge \mu_L(z_3) \wedge \mu_L(z_4) \wedge \mu_S(z_5) \wedge \mu_L(z_6) \wedge \mu_L(z_7) \vee \\ &\mu_{Ln}(z_1) \wedge \mu_A(z_2) \wedge \mu_L(z_3) \wedge \mu_L(z_4) \wedge \mu_S(z_5) \wedge \mu_L(z_6) \wedge \mu_L(z_7) \end{aligned} \tag{3.14}$$

$$\begin{aligned} &\mu_{BA}(Z) = \wedge \mu_{Ln}(z_1) \wedge \mu_A(z_2) \wedge \mu_A(z_3) \wedge \mu_L(z_4) \wedge \mu_S(z_5) \wedge \mu_L(z_6) \wedge \mu_H(z_7) \vee \\ &\mu_{LN}(Z_1) \wedge \mu_L(z_2) \wedge \mu_L(z_3) \wedge \mu_L(z_4) \wedge \mu_S(z_5) \wedge \mu_L(z_6) \wedge \mu_H(z_7) \vee \\ &\mu_A(z_1) \wedge \mu_L(z_2) \wedge \mu_A(z_3) \wedge \mu_L(z_4) \wedge \mu_S(z_5) \wedge \mu_L(z_6) \wedge \mu_H(X_7) \end{aligned} \tag{3.15}$$

$$\begin{aligned} &\mu_A(Z) = \wedge \mu_A(z_1) \wedge \mu_A(z_2) \wedge \mu_A(z_3) \wedge \mu_A(z_4) \wedge \mu_A(z_5) \wedge \mu_A(z_6) \wedge \mu_A(z_7) \vee \\ &\mu_A(Z_1) \wedge \mu_L(z_2) \wedge \mu_A(z_3) \wedge \mu_A(z_4) \wedge \mu_S(z_5) \wedge \mu_L(z_6) \wedge \mu_A(z_7) \vee \\ &\mu_{Ln}(z_1) \wedge \mu_L(z_2) \wedge \mu_A(z_3) \wedge \mu_L(z_4) \wedge \mu_S(z_5) \wedge \mu_L(z_6) \wedge \mu_H(X_7) \end{aligned} \tag{3.16}$$

$$\begin{aligned} &\mu_A(z_1) \wedge \mu_A(z_2) \wedge \mu_A(z_3) \wedge \mu_H(z_4) \wedge \mu_A(z_5) \wedge \mu_A(z_6) \wedge \mu_L(z_7) \vee \\ &\mu_S(Z_1) \wedge \mu_H(z_2) \wedge \mu_A(z_3) \wedge \mu_A(z_4) \wedge \mu_A(z_5) \wedge \mu_A(z_6) \wedge \mu_L(z_7) \end{aligned} \tag{3.17}$$

$$\begin{aligned} &\mu_H(X) = \wedge \mu_S(z_1) \wedge \mu_H(z_2) \wedge \mu_H(z_3) \wedge \mu_H(z_4) \wedge \mu_{Ln}(z_5) \wedge \mu_H(z_6) \wedge \mu_L(z_7) \vee \\ &\mu_S(Z_1) \wedge \mu_A(z_2) \wedge \mu_A(z_3) \wedge \mu_H(z_4) \wedge \mu_{Ln}(z_5) \wedge \mu_H(z_6) \wedge \mu_L(z_7) \vee \\ &\mu_S(z_1) \wedge \mu_H(z_2) \wedge \mu_H(z_3) \wedge \mu_H(z_4) \wedge \mu_{Ln}(z_5) \wedge \mu_A(z_6) \wedge \mu_L(X_7) \end{aligned} \tag{3.18}$$

The fuzzy logic confirmation technique allows observing an indicator predicted as fuzzy sets. Fuzzy sets estimate the environmental and economic performance of a project variant for the fixed vector of influencing factors. In order to move from the obtained fuzzy sets to quantity assessment, a dephasing procedure must be performed. It consists in the transformation of fuzzy information into a distinct form. Among various methods of defuzzification, the most common is finding the "center of gravity" of a flat figure, which is limited by the function of fuzzy set membership and horizontal coordinate. The fuzzy inference model, together with the dephasification procedure, provides an opportunity to monitor the changes in the baseline – the environmental and economic efficiency of the project.

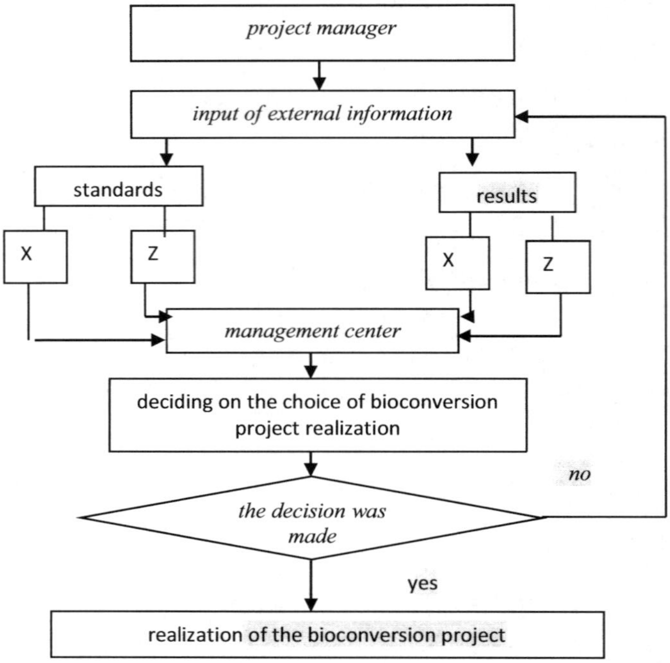

Figure 3.2 Structural and logical model of managing ecological and economic feasibility of an innovative bioconversion project.

In order to assess the economic efficiency of the mechanisms for providing thermal stabilization process in biogas installations at the conceptual phase of the life cycle of an investment project, a structural logical model for minimizing the bioconversion products cost by managing renewable energy sources has been offered (Figure 3.2). This model allows taking into account required variable data about the dynamic parameters of the thermal stabilization process in the production of biogas as ecological energy source.

3.3 CONCLUSIONS

- Formalization and hierarchical classification of the quantity and quality factors were completed, which are important and significantly help determine innovative appeal of a bioconversion project variant.
- Models of managing the assessment of bioconversion ecological-economic efficiency based on fuzzy logic at ecological and economic levels were developed.
- For the first time, a structural and logical model for managing ecological and economic feasibility of an innovative bioconversion project was created.

REFERENCES

Drukovani, M.F., Yaremchuk, O.S. & Sosnovskaya, L.V. 2006. *Biomass Processing Technology: Monograph*. Vinnytsia: VNAU.

Geletukha, G.G. & Martseniuk, S.A. 1999. A review of biogas production and utilization technologies in landfills and landfills of solid household waste and prospects for their development in Ukraine. *Ecotechnology and Resource Saving* 4: 6–14.

Kukharchuk, V.V. Kazyv, S.S. & Bykovsky, S.A. 2017. Discrete wavelet transformation in spectral analysis of vibration processes at hydropower units. *Przeglad Elektrotechniczny* 93(5): 65–68.

Noyola, A., Morgan-Sagastume, J. & Lopez-Hernandez, J. 2006. Treatment of biogas produced in anaerobic reactors for domestic wastewater: odor control and energy/recourse recovery. *Environmental Science and Biotechnology*: 93–114.

Ratushnyak, G.S. & Anokhina, K.W. 2013. *Energy Efficient Bioconversion Technologies and Equipment: A Monograph*, Vynnitsia: Vinnytsia National Tech. Univ., VNTU, 160 p.

Redko, A.O., Bezrodnyi, M.K., Zagoruchenko, M.V. & Dolinsky, A.A. 2016. *Low Potential Energy: A Textbook*. Kharkiv: Madrid Printing House.

Rotshtein, A.P. 1999. *Intelligent Identification Technologies. Fuzzy Sets, Genetic Algorithms, Neural Networks*. Vinnitsa: Universum.

Rotshtein, A.P, Larushkin, E.P. & Katelnikov, D.I. 1997 Multivariate analysis of bioconversion technological process on the basis of linguistic information. *Bulletin of VPI* 3: 38–45.

Rotshtein, A.P., Larushkin, E.P. & Mityushkin, Y.I. 2008. *Soft Computing in Biotechnology: Multi Factorial Analysis and Diagnostics: Monograph*, Vynnitsia: Vinnytsia National Tech. Univ., VNTU, 144 p.

Suresh, B., Baral, S., Pudasaini, S.P., Khanal, S.N., Gurung & D.B. 2013. Mathematical modelling, finite element simulation and experimental validation of biogas–digester slurry temperature. *International Journal of Energy and Power Engineering* 2(2): 128–135.

Weiland, P. 2003. Production and energetic use of biogas from energy crops and wastes in Germany. *Applied Biochemistry and Biotechnology* 109: 263–274.

Zabarnyi, G.M. & Shurchkov, A.V. 2002. *Energy Potential of Non-Traditional Energy Sources of Ukraine*. Kyiv: Institute of Engineering Thermophysics of NAS of Ukraine, 210 p..

Zadeh, L.A. 1975. The concept of a linguistic variable and its application to approximate reasoning, *Journal of Information Science*, 8: 199–249.

Zhelikh, V.M., Lesik, H.R. & Piznak, B.I. 2013. Investigation of the exergy efficiency of low-temperature solar collectors, *Modern Technologies, Materials and Structures in Construction* 1: 135–142.

Zuev, O.O. 2009. Economic aspects of introduction of modern biogas installations. *Proceedings of the TDATU* 9(5): 88–92.

Chapter 4

Increasing the Efficiency of Municipal Solid Waste Pre-processing Technology to Reduce Its Water Permeability

O. V. Bereziuk, M. S. Lemeshev, and Volodymyr V. Bogachuk
Vinnytsia National Technical University

Roman B. Akselrod
Kyiv National University of Construction and Architecture

Alla P. Vinnichuk
Vinnytsia Mykhailo Kotsiubynskyi State Pedagogical University
Kyiv National University of Construction and Architecture

Andrzej Smolarz
Lublin University of Technology

Mukaddas Arshidinova
Al-Farabi Kazakh National University

Olena Kulakova
Satbayev University

CONTENTS

4.1 INTRODUCTION

Annually, more than 54 million municipal solid wastes (MSW) are formed in the settlements of Ukraine. The main part of MSW is buried in 4,530 landfills and landfill area of almost 7,700 ha and only partially recycled or disposed at waste incineration plants, in contrast to the European Union countries, where modern technologies are widely introduced with MSW management (Moroz et al., 2003). During the 1999–2014, the total area of the landfills in Ukraine increased three times. The area of

DOI: 10.1201/9781003177593-4

overloaded landfills increased almost twice and the landfills that violate environmental safety standards, including pollution of soil and groundwater filtration, increased more than in 3.1 times. This poses a threat to life and health of the population, especially the nearest inhabited localities. MSW landfill filtrate refers to highly contaminated wastewater and is the main factor in the negative influence of landfills on the environment due to the presence of toxic substances, heavy metals (lead, chromium, cadmium, mercury), organic, inorganic compounds, etc. that are part of the waste or products of their decomposition (Popovych et al., 2018). A significant distinction of the filtrate from other types of wastewater is their uneven accumulation during the year due to seasonal fluctuations in the level of precipitation (Lototskyi & Bistrom, 2005). According to the literature data (Shyshkyn, 2007), depending on the climatic conditions, the volume of filtrate, formed over a year from the area of the body of MSW landfill 1 ha, averages 2,000–4,000 m^3/year. The formation of the filtrate contributes to a significant relative humidity of MSW: for mixed MSW, it is within 39%–53%, for the food fraction of MSW in the spring and summer period it is 60%–64%, and in autumn, it is 75%–92% (Varnavskaia, 2008). Moreover, the moisture emitted from the MSW thickness as a result of biochemical processes is accompanied by the formation of liquids in anaerobic decomposition of their organic component. As a result of the pressure placed above the layers of MSW, as well as under the action of gravity, moisture is pressed, and the landfill is formed into a kind of aquifer. The formed filtrate accumulates at the bottom of the pit of the landfill under a layer of MSW, which virtually eliminates its natural evaporation. The main MSW fraction is organics (biomass), which causes biological pollution of the environment by the helminths and pathogenic microorganisms. No normative document of Ukraine on maintenance and operation of MSW landfills provides a ban on the accumulation of filtrate (Podchashynskyi et al., 2017). The violation of environmental legislation is only the fact of hit of poisonous substance in the environment outside the landfill. Therefore, the measures to reduce the amount of filters by reducing water permeability should be developed to prevent harmful substances from infiltrating into the environment. The higher the chemical heterogeneity of MSW, the greater the danger they pose to humans and the environment (Lototskyi & Bistrom, 2005). One way to change the properties of MSW is their initial processing, which can include dewatering, shredding, compaction, sorting, etc. In our opinion, in order to increase efficiency, primary recycling of MSW must be performed at the stage of their loading in special vehicles (body dustcarts), which are used to collect and transport waste to the places of disposal and incineration. Annually in Ukraine, more than 45,000 tons of fuel is spent for the transportation of 4,000 MSW dustcarts to the place of disposal outside the sanitary zone of 30 km. Deterioration of the fleet of dustcarts of Ukrainian municipal enterprises reaches almost 70%.

According to the decree of the Cabinet of Ministers of Ukraine No. 265 (Kabinet Ministriv Ukrainy, 2004; Pilov, 2012), the application of modern highly effective dustcarts in the communal economy of the country, as the main link in the structure of machines for the collection and primary processing of MSW, is a relevant scientific and technical problem. Improving the efficiency of the primary recycling technology for MSW to reduce water permeability, aimed at minimizing their negative environmental impact and improving environmental safety in general, is one of the important tasks for solving this problem.

The purpose of the study is to increase the efficiency of the primary processing technology of MSW to reduce the water permeability, aimed at minimizing their negative environmental impact and improving the environmental safety as a whole.

The implementation of an innovative scenario for the development of the MSW management sector in Ukraine is a necessary prerequisite for effective waste management (Berezyuk et al., 2019). The analysis of the literature sources establishes that the increase in the number of small particles of waste contributes to reducing the hydraulic conductivity of MSW at the landfills (Reddy et al., 2009). According to the authors of the work (Gavelytė et al., 2015) reduction of the MSW particles size from 50 to 10 mm decreases their permeability by almost 22 times when buried on a proving ground, which contributes to reducing the intensity of soil pollution and groundwater filtration, as evidenced by the data in Table 4.1.

MSW contains ferrous and non-ferrous metals that are able to corrode, participate in the oxidationreduction reactions, to form complex compounds with organic ligands – the products of biochemical decomposition of an organic part of MSW, to form the hard-soluble hydroxide, carbonates, phosphates, and sulfides, which with filtration can infiltrate to soils and groundwater adjacent to landfills (Kulchytska-Zhyhailo, 2008). In the article by Pyrskyi et al. (2009), the basic provisions of the method were developed, which provide the possibility to detect and calculate the leakage amounts of filtration from the MSW polygons based on balance calculations and the results of instrumental observations without the use of direct measurements. In the work by Abu Qdais and Alsheraideh (2008), it is recommended to consider the size of their particles during the investigations at MSW landfills. MSW shredding increases the surface area for the microbial activity during the decomposition of organic materials (Agunwamba, 2001). After reducing the size of MSW particles from 50.8 to 12.7 mm, their density is increased by 3.7%–6.8%, which is explained by the potential influence of previously clogged internal voids and access to them due to the reduction of the size of the separate components of the waste parts (Yesiller et al., 2014). Thus, the analysis of the literature sources dealing with the treatment of MSW confirms the need for their shredding, which contributes to reducing the intensity of soil pollution and groundwater filtration and increases the environmental safety in general.

Table 4.1 Water permeability of various sizes of MSW particles (Gavelytė et al., 2015)

Size of MSW particles, [mm]	*Water permeability of waste [mm/h]*
≈0	≈0
5	0.0753
10	0.58
20	1.91
40	4.73
50	8.34
60	12.7
80	21.3
100	32.8

In the article by Bereziuk (2019), the adequate quadratic regression dependence of such indicators of the shredding process as the residue on the sieve and energy capacity for shredding of MSW from the basic parameters of influence: the average size of particles, the primary relative density of MSW, and the diameter of the auger screw on the last coil. In the work by Podchashynskyi et al. (2017), the hydrochemical composition of the filtration was investigated and analyzed on the example of MSW landfill in Zhytomyr (Ukraine); the experimental data and the revealed features of them are generalized in the form of mathematical models of the dynamics of changes of chemical components in the filtrate samples, which allow predicting the composition of the filtrate measures for its decontamination. The regression analysis presented in Shpakova (2018) showed that the logarithmic model is the best in describing the relation of production and consumer waste generation for the regions of the Far Eastern Economic Region of Russia and the Region as a whole. However, the specific mathematical dependences of filtration on MSW particle size, as a result of analysis of known publications were not revealed by the authors.

4.2 METHODS

The following methods were used for research and analysis: analysis of literature sources, regression analysis, and computer data processing.

Regressions were performed on the basis of linearization transformations, which allow reducing nonlinear dependence to linear. Coefficients of the regression equations were carried out by the least-squares method with the help of a developed computer program "RegAnaliz", which is protected by the certificate on the registration of copyright in the work, and described in the work (Bereziuk, 2014). The block scheme of the program is shown in Figure 4.1.

The "RegAnaliz" program allows for regression analysis of the results of one-factor experiments and other pair dependencies with the choice of the best type of function from 16, the most common variants according to the maximum correlation coefficient of saving the results in the MS Excel and Bitmap formats.

4.3 MAIN RESULTS OF THE RESEARCH

The results of the regression analysis are given in the Table 4.2, where the grey color is indicated by a cell with the maximum correlation coefficient R for paired regression.

Therefore, according to the results of the regression analysis, based on data given in Table 4.1, the following regression dependence between the water permeability and the size of MSW particles was finally adopted as the most adequate:

$$k_f = 0.0002625 + 0.004484 d^{1.932} \, (\mathrm{mm/h}) \tag{4.1}$$

where k_f is the coefficient of permeability, mm/h, and d is the size of MSW particles, mm.

In Figure 4.2, the graphical dependence of the water permeability on the MSW particle size, constructed with regression (4.1), confirming the previously determined high accuracy of received theoretical dependence in comparison with experimental data, is obtained by the authors (Gavelytė et al., 2015).

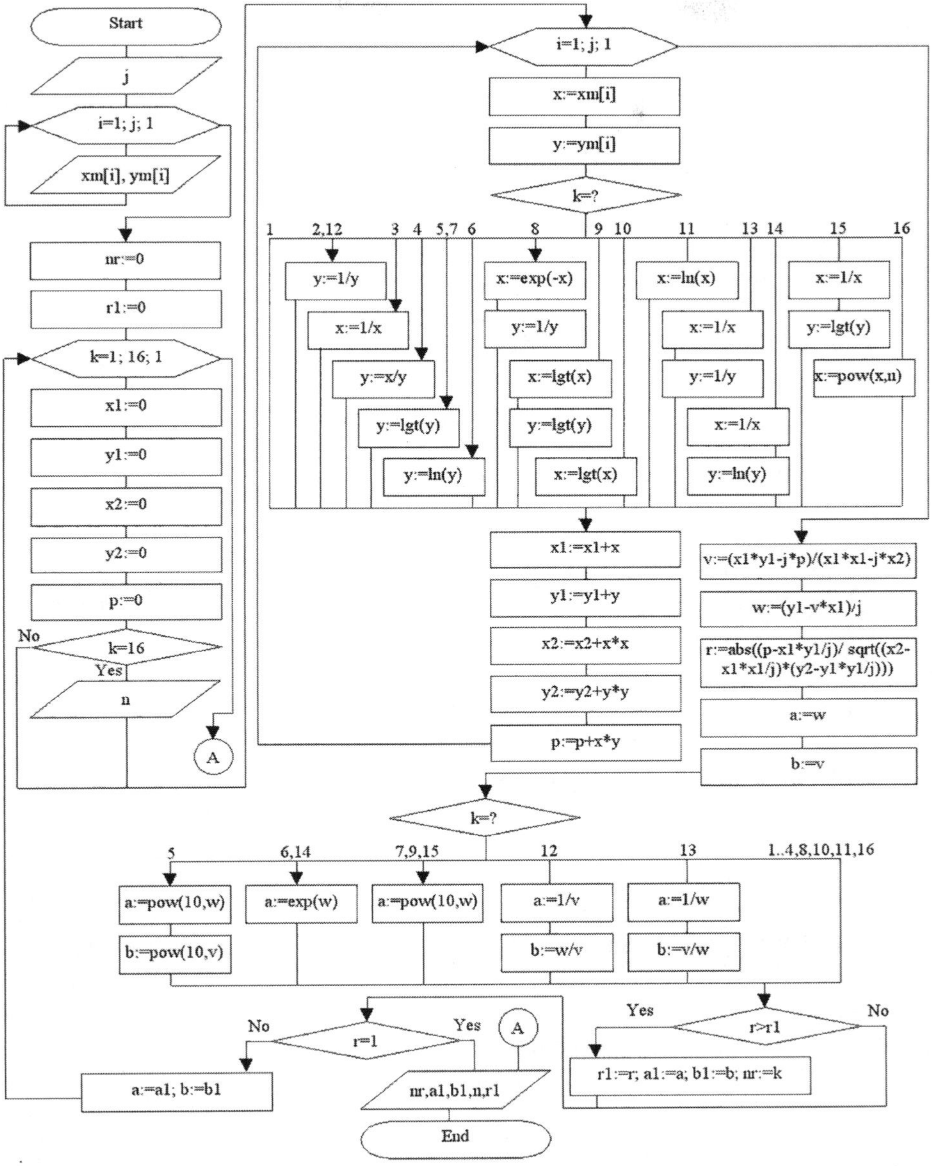

Figure 4.1 Block scheme of the "RegAnaliz" program algorithm.

The necessary coefficient of water permeability of soils in the areas of MSW disposal, providing normative indicators of hydrogeological environment, is 1×10^{-9} m/s (DBN B.2.4-2:2005) , which is equal to 0.0036 mm/h for which special measures on an arrangement of hydro-protective screens are not required. This value approaches the natural state of the clay and heavy loam, regarded as the most promising soils at the basis of the MSW landfill (DSanPiN, 1999, Orlova, 2004). Then, based on the

Table 4.2 Results of regression analysis

Type of regression	*Correlation coefficient R*	*Type of regression*	*Correlation coefficient R*
$y = a + bx$	0.95919	$y = ax^b$	0.99900
$y = 1/(a + bx)$	0.44054	$y = a + b \cdot \lg x$	0.30346
$y = a + b/x$	0.30257	$y = a + b \cdot \ln x$	0.30475
$y = x/(a + bx)$	0.45732	$y = a/(b + x)$	0.44054
$y = ab^x$	0.43607	$y = ax/(b + x)$	0.99892
$y = ae^{bx}$	0.43873	$y = ae^{b/x}$	0.99898
$y = a \cdot 10^{bx}$	0.43607	$y = a \cdot\cdot 10^{b/x}$	0.99900
$y = 1/(a + be^{-x})$	0.99898	$y = a + bx^{n+2}$	0.9993 9

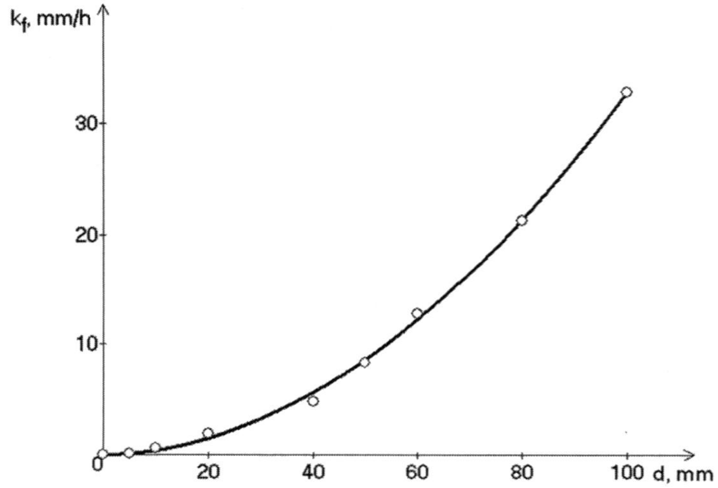

Figure 4.2 Dependence of water permeability on the size of MSW particles: the measured O, theoretical —.

dependencies (4.1), it is possible to obtain an expression to determine the necessary power of MSW shredding to ensure regulatory filtration coefficient at their burial at the proving ground

$$d_{\min} = (223k_f - 0.05854)^{1/1.932}\ (\text{mm}) \tag{4.2}$$

Substituting in dependence (4.2) the value of regulatory filtration coefficient at burial on proving ground, we obtain the necessary minimum size of MSW particles after their primary processing by shredding

$$d_{\min} = (223 \cdot 0.0036 - 0.05854)^{1/1.932} = 0.858\ (\text{mm})$$

Table 4.3 Indicative content of components by the size of particles of MSW fractions in Ukraine, % by weight (Cherp & Vynychenko, 1996)

Component	*Content, % of constituents by size of particles of MSW fractions, mm*				
	>250	*150–250*	*100–150*	*50–100*	*<50*
Food waste	–	0–1	2–10	7-2.16	17–21
Paper, cardboard	3–8	8–10	9–11	7–8	2–5
Tree	0.5	0–0.5	0–0.5	0.5	0–0.5
Metal	–	0–1	0.5–1	0.8–1.6	0.3–0.5
Textile	0.2–1.3	1–1.5	0.5–1	0.3–0.8	0–0.6
Bone	–	–	–	0.3–0.5	0.5–0.9
Glass	–	0–0.3	0.3–1	1–2	1–1.6
Leather, rubber	–	0–1	0.5–2	0.5–1.5	–
Stones	–	–	0.2–1	0.5–1.8	0.5–2
Plastics	0–0.2	0.5–1	1–2.2	1–2.5	0.2–0.5
Other	0–0.3	0.2–0.6	0–0.5	0–0.4	0–0.5
Screenings < 16 mm	–	–	–	–	4–6
Total	7.0	13.3	22.1	25.3	32.3

In order to determine the need to grind the MSW before burial on the proving ground, the obtained value of the required minimum particle size of 0.858 mm was compared with the approximate dimensions of the particles of the MSW fractions in Ukraine, given in the work by Cherp and Vynychenko (1996).

As seen in Table 4.3, the majority of MSW have a particle size greater than 0.858 mm as the fractional composition, and in general, which confirms the need to shred them before dumping at the landfills to reduce the water permeability of the filtration, and, consequently, to reduce the environmental pollution. However, the shredding processes are the most costly and power-consuming processes in the general scheme of MSW processing. Therefore, waste shredding can be realized during dewatering and pre-compaction MSW by using a conical screw, experimental research of which is given in Pilov (2012). At the same time, the energy intensity of MSW shredding process was 258.3–389.1 kWh/t, and the weighted average size of the particles of crushed MSW (module grinding) amounted to 9.2–10.95 mm, which ensures a decrease in water permeability of the waste according to equation 4.1 to 0.33–0.46 mm/h, respectively, which corresponds to $9.2 \cdot 10^{-8} – 1.28\ 10^{-7}$ m/s, reducing the intensity of environmental impact. However, as the process of MSW shredding is proposed to be carried out during their dehydration and pre-compaction, rather than a separate technological operation, that is, without the cost of additional energy, it increases the efficiency of pre-processing waste technology as a whole.

Regressive dependencies (4.1) and (4.2) may be used to create a methodology for the engineering calculations of the equipment parameters for shredding MSW as one of the methods of their primary processing. Moreover, these dependencies can be taken into account in the development of a strategy for MSW management, aimed at minimizing their negative impact on the environment and improving the environmental safety as a whole.

4.4 CONCLUSIONS

Regression dependence of water permeability on the particle size of MSW, which can be used during the creation of the technique of engineering calculation of parameters of equipment for shredding of MSWs as one of the methods of their primary processing, is offered. Additionally, this dependence can be taken into account during the development of the MSW management strategy aimed at minimizing their negative impact on the environment and improving environmental safety as a whole.

The required minimum size of particles of MSW after their primary processing is determined by crushing and amounts to 0.858 mm, at which the normative coefficient of permeability is ensured at proving ground 0.0036 mm/h, for which no need to foresee special measures on the arrangement of hydro-protective screens.

It was established that during dehydration and pre-compaction of MSW, the power supply of their grinding process was 258.3–389.1 kWh/t, and the weighted average size of the crushed waste particles amounted to 9.2–10.95 mm, which will ensure reducing water permeability to 0.33–0.46 mm/h, respectively (which corresponds to $9.2\ 10^{-8}$–$1.28\ 10^{-7}$ m/s), reducing the influence on the environment. However, because the process of shredding MSW is proposed to carry out during their dewatering and pre-compaction, rather than a separate technological operation and the supply of additional energy is not needed, thus the increased efficiency of whole the MSW pre-processing is reached.

REFERENCES

Abu Qdais, H. & Alsheraideh, A. 2008. Kinetics of solid waste biodegradation in laboratory lysimeters. *Jordan Journal of Civil Engineering* 2(1): 45–52.

Agunwamba, J.C. 2001. *Waste: Engineering and Management Tools*. Enugu: Immaculate Publications Ltd.

Bereziuk, O.V. 2014. Vstanovlennia rehresii parametriv zakhoronennia vidkhodiv ta potreby v ushchilniuvalnykh mashynakh na osnovi kompiuternoi prohramy "RegAnaliz". *Visnyk Vinnytskoho politekhnichnoho instytutu* 1: 40–45.

Bereziuk, O.V. 2019. Eksperymentalne doslidzhennia protsesu podribnennia tverdykh pobutovykh vidkhodiv pid chas znevodnennia shnekovym presom. *Visnyk Vinnytskoho politekhnichnoho instytutu* 5: 75–80.

Berezyuk, S., Tokarchuk, D. & Pryshliak, N. 2019. Resource potential of waste usage as a component of environmental and energy safety of the sate. *Journal of Environmental Management and Tourism* 10(5):1157–1167.

Cherp, O.M. & Vynychenko, V. N. 1996. *Problema tverdykh bytovykh otkhodov: kompleksnyi podkhod* (*The Problem of Municipal Solid Waste: An Integrated Approach*). Moskva: Ekolain.

DBN B.2.4-2. 2005. Полігони твердих побутових відходів. Основні положення проектування (Landfills for solid waste. Basic design provisions).

DSanPiN 2.2.7.029-99. 1991. Derzhavni sanitarni pravyla ta normy. Hihiyenichni vymohy shchodo povodzhennya z promyslovymy vidkhodamy ta vyznachennya yikh klasu nebezpeky dlya zdorov'ya naselennya (Hygienic requirements for waste management and determination of their hazard class for public health).

Gavelytė, S., Bazienė, K., Woodman & N., Stringfellow, A. 2015. Permeability of different size waste particles. *Science – Future of Lithuania* 7(4): 407–412.

Kabinet Ministriv Ukrainy. 2004. *Postanova No. 265 "Pro zatverdzhennia Prohramy povodzhennia z tverdymy pobutovymy vidkhodamy"*. http://zakon1.rada.gov.ua/laws/show/265-2004-%D0%BF.

Kulchytska-Zhyhailo, L. 2008. Standarty YeS ta chynni v Ukraini normy i pravyla proektuvannia ta ekspluatatsii polihoniv tverdykh pobutovykh vidkhodiv. *Materialy nauk.-tekhn. konf. "Polihony tverdykh pobutovykh vidkhodiv: proektuvannia ta ekspluatatsiia, vymohy Yevropeiskoho Soiuzu, Kiotskyi protokol"*, Lviv, Ukraina, 2008, pp. 145–155.

Lototskyi, O.B. & Bistrom, Y. 2005. Natsionalna stratehiia povodzhennia z tverdymy pobutovymy vidkhodamy v Ukraini – shliakhy do stabilnoho maibutnoho. *Sbornyk dokladov mezhdunarodnoho konhressa "ETEVK-2005"* Yalta, May 24–27, pp. 47–51.

Moroz, O.V., Sventukh, A.O. & Sventukh, O.T. 2003. *Ekonomichni aspekty vyrishennia ekolohichnykh problem utyli-zatsii tverdykh pobutovykh vidkhodiv: monohrafiia*. Vinnytsia: Universum.

Orlova, T.A. 2004. Problema otsenky pryhodnosty terrytoryi dlia razmeshchenyia polyhonov TBO. *Mistobuduvannia ta terytorialne planuvannia* 17: 211–215.

Pilov, P.I. 2012. *Obgruntuvannia parametriv obladnannia tekhnolohichnykh linii pererobky promyslovykh ta tverdykh pobutovykh vidkhodiv z otrymanniam produktiv dlia podalshoho vykorystannia (zvit, zakliuchnyi) tema HP-447.* Dnipropetrovsk: 107 c.

Podchashynskyi, I.O., Kotsiuba, I.H. & Yelnikova, T.O. 2017. Matematychne modeliuvannia i analiz vplyvu filtratu polihonu tverdykh pobutovykh vidkhodiv na dovkillia. *Vostochno-evropeiskyi zhurnal peredovykh tekhnolohyi* 10 (85): 4–10.

Popovych, V., Stepova, K. & Prydatko, O. 2018. Environmental hazard of Novoyavorivsk municipal landfill. *MATEC Web of Conferences* 247(00025): 1–8.

Pyrskyi, O.A., Olevska, T.V. & Kolunaiev, Y.V. 2009. Vyznachennia naiavnosti vytokiv filtratu z polihoniv tverdykh pobutovykh vidkhodiv nepriamym metodom. *Visnyk NTUU "KPI". Seriia "Hirnytstvo"* 18: 117–123.

Reddy, K.R., Hettiarachchi, H. & Parakalla, N., Gangathulasi, J., Bogner, J., Lagier, T. 2009. Hydraulic conductivity of MSW in landfills. *Journal of Environmental Engineering* 135(8): 677–683.

Shpakova, R.N. 2018. Study of the dependency between the gross regional product and the production and consumer waste generation levels. Smart technologies and innovations in design for control of technological processes and objects: Economy and production. FarEastCon 2018. *Smart Innovation, Systems and Technologies* 139: 34–41.

Shyshkyn, I.S. 2007. *Snyzhenye ekolohycheskoi nahruzky polyhonov TBO na obekty hydrosfery na zavershaiushchykh etapakh zhyznennoho tsykla: avtoref. dys. kand. tekhn. Nauk*. Perm: Perm. hos. tekhn. un-t.

Varnavskaia, Y.V. 2008. Analyz uslovyi obrazovanyia y sostava stochnykh vod polyhonov tverdykh bytovykh otkhodov. Ekolohyia y promyshlennost 1: 39–43.

Yesiller, N., Hanson, J.L., Cox, J.T. & Noce D.E. 2014. Determination of specific gravity of municipal solid waste. *Waste Management* 34(5): 848–858.

Chapter 5

Assessment of Pesticide Phytotoxicity with the Bioindication Method

Roman V. Petruk, Natalia M. Kravets, and Serhii M. Kvaterniuk
Vinnytsia National Technical University

Yuriy M. Furman
Vinnytsia Mikhailo Kotsiubynskyi State Pedagogical University

Róża Dzierżak
Lublin University of Technology

Mukaddas Arshidinova
Al-Farabi Kazakh National University

Assel Jaxylykova
Al-Farabi Kazakh National University
Institute of Information and Computational Technologies CS MES RK

CONTENTS

5.1 INTRODUCTION

The online information on the phytotoxicity of contaminated water can be obtained using test objects (seeds and seedlings of plants) and various test indicators (dynamics of seed germination, percentage of germination, length of main and lateral roots, the height of shoot, etc.). Carrying out the experiments on the influence of various man-made substrates on plant objects under controlled conditions allows solving many problems: determining the causes of different plant resistance and the tendency to adapt to toxicants, detecting the influence of a specific environmental factor, excluding

DOI: 10.1201/9781003177593-5

the effect of other factors, and identifying the lethal dose of the pollutant (Goncharuk et al. 2012; Tanner et al. 2002; Kovalenko et al. 2016).

5.2 PROBLEM STATEMENT

Today, the problem of contamination of natural objects with technogenic waste is exacerbated. Industrial wastes poison the air, water and soil, as well as adversely affect living organisms and are toxic to them. A major environmental problem that requires immediate resolution is the pollution of water and soil by oil and petroleum products, heavy metals, pesticides, and other substances. Physical and chemical methods have traditionally been used to neutralize them. However, every year there is a growing interest in using biotechnological methods of waste disposal and environmental clean-up as more efficient and economical, and most importantly, environmental-friendly cleaning methods. The most common methods of water, air, and soil purification are the adsorption methods. However, most known sorbents have common disadvantages (high cost, low sorption capacity, etc.). The production of sorbents in the traditional way is characterized by multistage, complexity of the used equipment, limited raw material base, and the like pollutant (Goncharuk et al. 2012).

Considerable attention is paid to new, highly efficient technologies that are based on the application of biosorbents, which combine the benefits of sorption and biodestructive methods of eliminating pollution. Biodegradable sorbents locate contamination and destroy the adsorbed products with biological objects of different levels of organization (microorganisms, algae, plants). This achieves effective pollution clearance.

Bioindication allows evaluating (Tanner et al. 2002):

- complex, integral influence of pollutants on the species composition and quantity of hydrobionts, characterizing the quality of waters as their habitat;
- changes in water quality over a short period of time;
- water quality in view of its suitability for business and human needs.

Biotesting, as a method of assessing the state of the aquatic environment, is used:

- in determining the phytotoxicity of water;
- for establishing the effects of xenobiotics aftereffects in the aquatic environment;
- during the toxicological assessment of sewage (industrial, domestic, agricultural, drainage), polluted natural waters in order to identify the potential sources of pollution;
- during ecological examination of new materials and substances.

In recent decades, biotesting has become a widely recognized and mandatory element of the toxic pollution control system in many countries (Kovalenko et al. 2016).

The choice of bioindicator species among the aquatic organisms inhabiting the reservoir implies that they possess the following characteristics: high ecological accuracy of the bioindicator response to the change of the environmental factor being indicated; relatively high number of indicator species; widespread in the ecosystem; simplicity in determining the taxonomic identity; and availability of the information on the ecology of the species.

Microorganisms contribute greatly to cleansing the environment from xenobiotics: bacteria, fungi, and microscopic algae that live in soil, freshwater bodies of water and seawater. For example, Aspergillum micromycetes contain up to 0.3% copper – 30,000 times more than in the environment. Many microorganisms accumulate uranium in large quantities: freshwater microalgae of chlorella – up to 0.4% of dry weight, actinomycetes – up to 4.5%, denitrifying bacteria – 14%, and specially selected cultures of yeast or pseudomonades – up to 50% (Martsenyuk et al. 2016; Goncharuk et al. 2016; Malik et al. 2014; Kusui 2002).

Two pesticide preparations were selected for the study, which are imported to the territory of Ukraine in the largest quantities: acetochlor (4,000 tons) and glyphosate (more than 3,000 tons).

Acetochlor is a colorless viscous liquid. Technical product is a dark brown liquid with a faint sweet odor. It is well soluble in organic solvents, poorly in water. According to other sources, the chemically pure preparation is an oily liquid of dark yellow color. It is well soluble in benzene, acetone, chloroform, diethyl ether, toluene, n-hexane, ethyl acetate, and ethanol as well as stable during storage. It constitutes a selective pre-eastern herbicide for the control of annuals of one- and two-weed weeds in cabbage, soybeans, corn, sugar beets, sunflower, onion, citrus, and fruit crops, in vineyards. It is an inhibitor of protein synthesis. It is adsorbed mainly by the roots of the plants. It is allowed to control annual weeds and dicotyledonous weeds in the crops of corn and soybeans, spraying the soil before sowing (with laying) until the emergence of crops.

Acetochlor, like other chloracetamides, is an inhibitor of seedlings and inhibits the processes of cellular respiration in plant roots. After application to the soil, the substance remains in its upper layer, penetrating the trash plants through the roots and seedlings. Herbicide suppresses the protein synthesis in sensitive plants. Since it does not affect weeds that have already descended, it should be applied before their emergence. The drug remains active for more than 12 weeks. No further herbicide treatments are needed at this time. Green corn forage, sunflower, and corn for grain can be used 70 days after the treatment.

Glyphosate is one of the most widely used non-selective systemic herbicides in the world. The glyphosate-based formulations are effective for destroying deep-rooted perennial weeds, annual and biennial broadleaf weeds, and water weeds. Glyphosate is used on a wide range of crops, in gardens and parks, in water and forestry pollutant (Goncharuk et al. 2012; Tanner et al. 2002; Guzzella et al. 2004). Glyphosate can be used at different stages of growth: before and after the seedling of crops (as herbicide) and before harvest (as desiccant). Glyphosate is one of the most common active substances in Ukraine, the EU and the world, used in herbicide preparations for agriculture. About 10,000 different drugs are officially registered in Ukraine. According to the State Register of Pesticides and Agrochemicals Permitted for Use in Ukraine, glyphosate is included in 137 preparations with valid registration.

Bioindication involves operational monitoring of the environment based on the observations of the state and behavior of biological objects (plants, animals, etc.).

This method is becoming increasingly popular as indicator plants have the following advantages:

- they summarize biologically important data on the environment;
- they are capable of responding to short-term and volatile emissions of toxicants;
- they respond to the speed of changes occurring in the environment;
- they indicate the places of accumulation of pollutants and the ways of their migration;
- they enable the development of estimates of the harmful effects of toxicants on humans and wildlife in the early stages and normalize the permissible load on ecosystems (Petruk et al. 2012).

A number of indicator plants respond to increased or decreased concentrations of micro- and macro-elements in the soil. This phenomenon is used for the preliminary assessment of soils, identification of possible sites for mineral exploration. One of the specific methods of monitoring environmental pollution is bioindication, determination of the degree of contamination of geophysical environments by living organisms, that is, bioindicators.

As pollutants, they use two types of pest control products that they produce, also in Ukraine. The information on the phytotoxicity of contaminated water can be obtained using test facilities (seeds and seedlings of plants) and various test indicators (dynamics of seed germination, percentage of germination, length of root and lateral roots, height of shoot, etc.). Carrying out the experiments on the influence of various man-made substrates on plant objects under controlled conditions allows solving many problems: determining the causes of different plant resistance and tendency to adapt to toxicants, detecting the influence of a specific environmental factor, excluding the effect of other factors, as well as identifying the lethal dose of the pollutant (Petruk et al. 2015, 2016).

5.3 EXPERIMENTAL RESEARCH

The experiment was conducted as follows: an appropriate solution of poison was added to the water sample, it was diluted in a ratio of 1:10. Because the wastewater is diluted from the natural in the specified ratio, that is, use the conditions as close to the natural state as possible. The seeds of wheat, barley, and corn were placed in Petri dishes, leveling the surface with a gauze disk, and then moistened with the same (10 ml) volume of the test solutions and the control sample (without the addition of poison). The cups were closed and kept for several days at room temperature for the diffusion of toxic substances into the water and for the germination of grains. The humidity of the substrates with seeds and poisonous component was in the range of 70%–80%. The control was the seeds without the addition of pest control, moistened to 70%–80% of full moisture capacity. The seeds were germinated at 23°C–25°C for 5 days. For more accurate data, three samples were used for each subject under study. The seeds of wheat, barley, and corn were selected because they respond well to the fluctuations in the content of the pollutant and is sensitive to the effects of toxic chemicals (Zabolotna et al. 2014; Wójcik & Smolarz 2017).

On the basis of the determination of the morphometric parameters of the test objects, it was found that the growth processes (Figures 5.1–5.3) of the studied seedlings were suppressed in all samples.

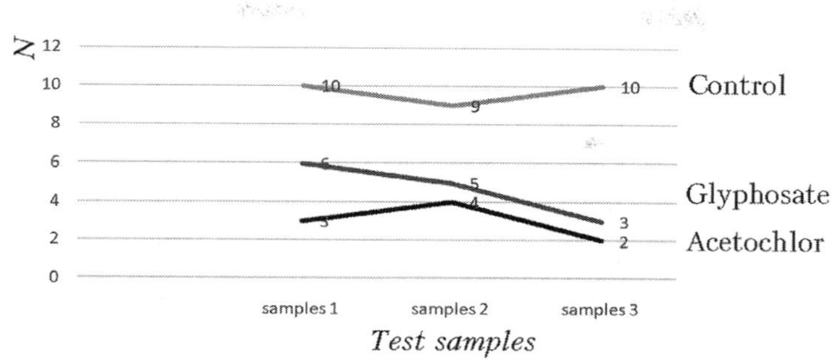

Figure 5.1 Intensity of germination of wheat grains in the samples tested.

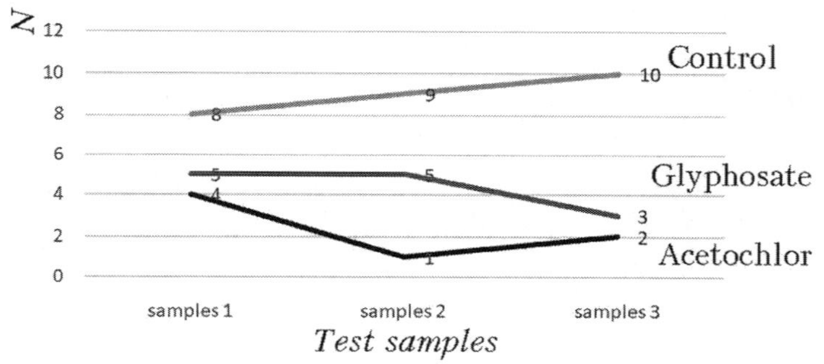

Figure 5.2 Intensity of germination of barley grains in the samples tested.

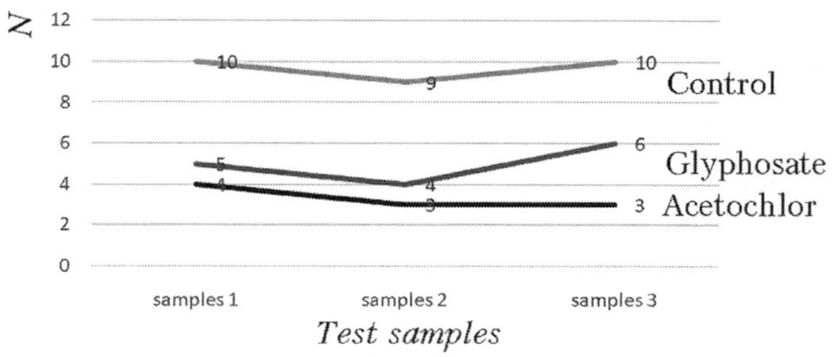

Figure 5.3 Intensity of germination of maize grains in the samples tested.

The following morphometric changes occurred in the studied samples: in the control sample (Figure 5.4), we observed practically complete germination of grains, with strong roots and stems; the average root length equals 54.6 cm. The following graphs show the average dynamics of grain germination and control samples (Figures 5.4–5.6).

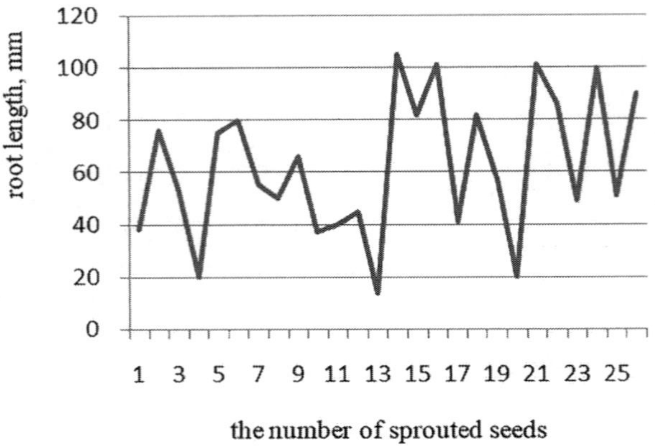

Figure 5.4 Dynamics of germination in the control sample.

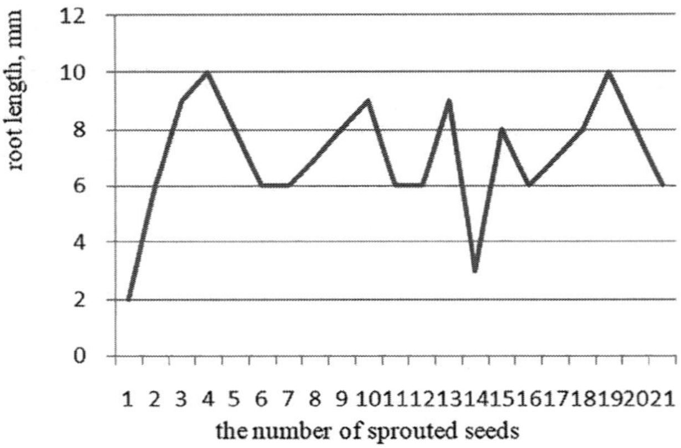

Figure 5.5 Dynamics of germination with the addition of glyphosate.

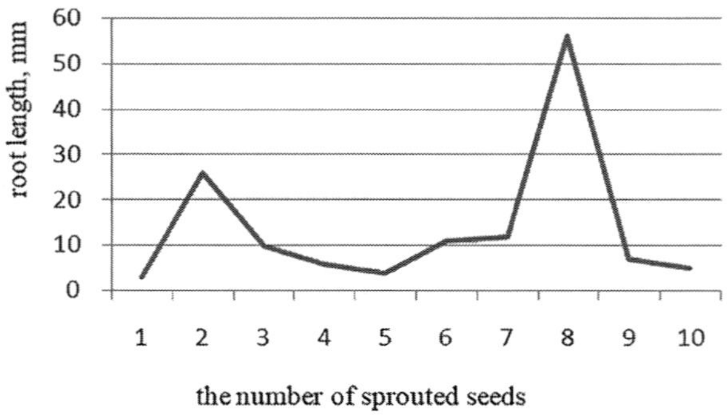

Figure 5.6 Dynamics of germination with the addition of acetochlorine.

At the same time, the worst dynamics of germination of grains, with the addition of the pesticide drug acetylchlor, the average root length is 4.8 cm, and the germination of grains decreased by 40% compared with the control sample.

It is known that the seeds of wheat, corn, and barley absorb well the solutions of toxic chemicals that in turn leads to the changes in metabolic reactions, resulting in that seeds may not sprout at all. From our experiments, it is seen that when the means of pest control on the seeds of radish sowing, less than half of the grains did not sprout at all.

The inhibition of the root growth processes of other test objects – wheat, barley, and maize – determined the toxicity level of the technogenic substrates being tested as medium and higher than average (Figure 5.7).

The phytotoxic effect for each sample (Table 5.1) of the studied object was calculated on the basis of the measurements performed.

The phytotoxic effect (*FE*, %) was determined as a percentage of the length of the root system according to formula 5.1:

$$FE = \frac{L_0 - L_x}{L_0} \cdot 100\% \qquad (5.1)$$

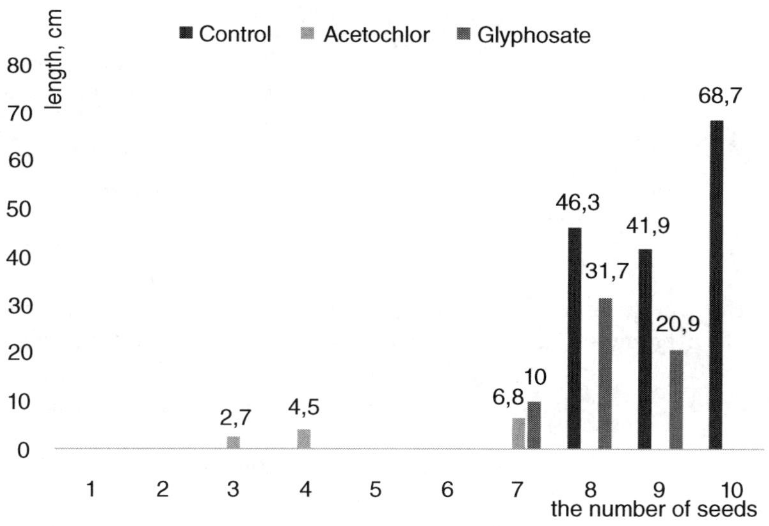

Figure 5.7 Dependence of grain growth on the introduced chemical.

Table 5.1 Phytotoxic effect for each sample

Pesticide preparation	*Phytotoxic effect, %*
Acetochlor	91
Glyphosate	91, 45

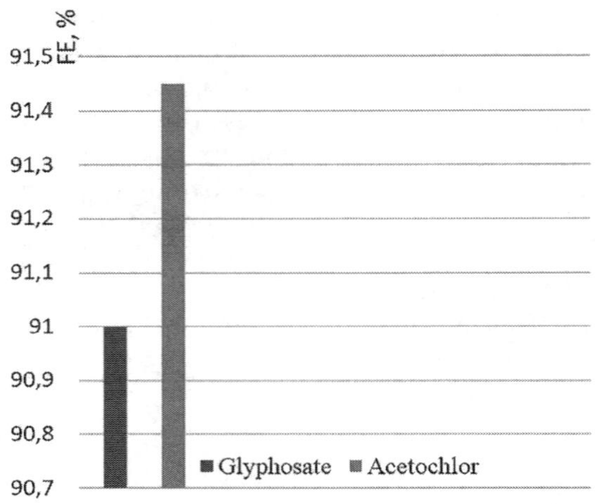

Figure 5.8 FE (%) of each toxic chemical.

Table 5.2 Scale of toxicity levels

Levels of inhibition of growth processes (phytotoxic effect), %	*Level of toxicity*
0–20	Absence or weak level
20.1–40	Average
40.1–60	Above average
60.1–80	High level
80.1–100	The maximum level

where

L_0 is the average root length of the plant grown on the control medium;

Lx is the average root length of a plant grown under the influence of a toxic factor.

The obtained data were processed with the methods of mathematical statistics (Figure 5.8).

The substance toxicity was assessed on a five-point scale (Table 5.2), which allows us to determine the phytotoxicity of test objects.

As a result, selected pesticide drugs are attributed to the maximum level of phytotoxicity. The main requirements that we were guided by in choosing this method of analysis are: expressiveness, accessibility and simplicity of the experiment.

5.4 CONCLUSIONS

The chapter analyzed the pollution of the aquatic environment, considered the classification of toxicants and also showed a detailed overview of the effects of pollutants on living aquatic components and human health. In addition, the methodology of biotesting and bioindication of water pollution was presented. A graphic depiction of the growth of the plant on the introduced toxic chemical, the dynamics of germination of grains in the control sample and with the addition of the appropriate pesticide preparation were shown.

Thus, according to the scale of toxicity levels (Table 5.2), it can be concluded that the phytotoxic effect of glyphosate and acetochlorine has the maximum level.

Therefore, having conducted the data of the study, it can be stated that there is a direct dependence between the inhibition of the morphometric parameters of the studied plants (radish sowing) and the introduced poison chemical. The greater the phytotoxic effect of the test sample, the smaller the length of the germinated roots and the greater the level of inhibition of plant development.

REFERENCES

Goncharuk, V., Kovalenko, V. & Zlatskii, I. 2012. Comparative analysis of drinking water quality of different origin based on the results of integrated bioassay. *Journal of Water Chemistry and Technology* 34(1): 61–64.

Goncharuk, V., Syroeshkin, A., Kovalenko, V., et al. 2016. Formation of a test systems and selection of test criteriain natural waters bioassay. *Journal of Water Chemistry and Technology* 38(6): 628-636.

Guzzella, L., Monarca, S., Zani, C., et al. 2004. In vitro potential genotoxic effects of surface drinking water treated with chlorine and alternative disinfectants. *Mutation Research* 564(2): 179–193.

Kovalenko, V., Zlatskii, I. & Goncharuk, V.J. 2016. Adaptive capabilities of hydrobionts to aqueous medium with different physicochemical parameters. *Journal of Water Chemistry and Technology* 38(1):56-61.

Kusui, T. 2002. Japanese application of bioassays for environ mental management. *Scientific World Journal* 2: 537–541.

Malik, A., Grohmann, E. & Akhtar, R. 2014. E*nvironmental Deterioration and Human Health: Natural and Anthropogenic Determinants*. Dordrecht, Springer: 8–25.

Martsenyuk, V., Petruk, V., Kvaternyuk, S., et al. 2016. Multispectral control of water bodies for biological diversity with the index of phytoplankton. *16th International Conference on Control, Automation and Systems (ICCAS 2016): ICCAS 2016 Conference Proceedings*: 988–993, Gyeongju, South Korea.

Petruk, R., Pohrebennyk, V., Kvaternyuk, S., et al. 2016. Multispectral television monitoring of contamination of water objects by using macrophyte-based bioindication. *16th International Multidisciplinary Scientific GeoConference SGEM 2016: SGEM2016 Conference Proceedings*; 5(2): 597–602, Albena.

Petruk, V., Kvaternyuk, S., Denysiuk, Y., et al. 2012. The spectral polarimetric control of phytoplankton in photobioreactor of the wastewater treatment. *Proceedings of SPIE* 8698: 86980H.

Petruk, V., Kvaternyuk, S., Kozachuk, A., et al. 2015. Multispectral televisional measuring control of the ecological state of waterbodies on the characteristics macrophytes. *Proceedings of SPIE* 9816: 98161Q.

Tanner, S., Baranov, V. & Bandura, D. 2002. Reaction cells and collision cells for ICP-MS: a tutorial review. *Spectrochimica Acta B*. 57: 1361–1452.

Wójcik, W. & Smolarz, A. 2017. *Information Technology in Medical Diagnostics*. London, Taylor & Francis Group CRC Press.

Zabolotna, N.I., Pavlov, S.V., Ushenko, A.G., Sobko, O.V. & Savich, V.O. 2014. Multivariate system of polarization tomography of biological crystals birefringence networks. *Proceedings of SPIE* 9166: 916615.

Chapter 6

Efficiency Assessment Functioning of Vibration Machines for Biomass Processing

Nataliia R. Veselovska, and Sergey A. Shargorodsky
Vinnytsia National Agrarian University

Larysa E. Nykyforova
National University of Life and Environmental Sciences of Ukraine

Zbigniew Omiotek
Lublin University of Technology

Imanbek Baglan and Mergul Kozhamberdiyeva
Al-Farabi Kazakh National University

CONTENTS

6.1 INTRODUCTION

The use of new energy-saving technologies has led to a significant development of the designs of vibrating machines and their widespread use, in particular for processing biomass. During their operation, the question of the efficiency and reliability of using this type of machine is quite relevant, due to the presence and possibility of using the reserves of its operation. The machines of this type must meet the requirements of quality and reliability in order to fulfill their official purpose.

The reliability and performance characteristics of vibrating machines are important technical and economic indicators related to the operation of systems for processing biomass. The increase of these characteristics opens the direction for the scientifically sound designation of reliability indicators, the achievement of these indicators in an economically optimal way. Improving the reliability and durability of vibrating machines has a serious reserve for saving money, materials, energy, and labor. To a large extent, the reliability and durability of a vibrating machine depend on extreme overloads. The qualification choice of materials and the correct calculations, taking into account the presence of a priori statistical information about the load at

DOI: 10.1201/9781003177593-6

the design stage, are the main sources of improving reliability without significantly raising the cost of the machine. Therefore, this research topic is relevant.

There are numerous known published monographs, textbooks, and periodic sources on the subject. The issues of ensuring the reliability of machines at the stages of design and operation are disclosed in the textbook (Sharov et al. 2015; Chernovol et al. 2010). An interconnected set of tasks is considered here: friction, aging, and wear. The causes of changes in the technical condition of machines and the physics of their failures were revealed. In the monograph (Iskovych-Lototsky et al. 2018), the approach for assessing the reliability of the effectiveness of ensuring the conditions of failure-free automated process was presented.

There are both fundamental and periodic sources where the results of the operation of vibratory-press equipment are published (Iskovych-Lototsky et al. 2017; Zhu et al. 2017). However, there are virtually no publications evaluating the efficiency and reliability of the operation of vibrating machines. In this regard, the topic of the article is relevant.

A research was conducted in order to propose and develop a system for evaluating the effectiveness and reliability of quantitative characteristics that is probabilistic-statistical in nature.

6.2 THE MAIN RESULTS OF THE STUDY

The problem of improving the reliability and efficiency of machines and structures constitutes an important technical and economic task, the solution of which opens the way for the science-based designation of the reliability indicators, achievement of these indicators in an economically optimal way. Improving the reliability and durability of machines represents a serious reserve for saving money, materials, energy, and labor costs. To a large extent, the reliability and durability of machines depend on current loads and actions. The correct choice of materials and the correction of calculations, taking into account a priori statistical information about the load at the design stage, are the main sources of improving reliability without significantly raising the cost of the machine.

The problem of the efficiency and reliability of using vibrating machines is associated with the presence and possibility of using the reserves of operation of the machine.

Therefore, to assess the effectiveness and reliability, it is necessary to introduce the quantitative characteristics that are probabilistic in nature. Since they can be determined not only experimentally but also by theoretical analysis, it is advisable to consider them from a statistical and probabilistic point of view.

As the quantitative characteristics of failure-free operation, the probability of the absence of failures, the frequency of failures, the failure rate, and the mean time between failures were used.

These questions are quite important in the direction of increasing the efficiency of technical diagnostics of the operation of vibrating machines (Sharov et al. 2015; Adamchuk et al. 2017; Vedmitskyi et al. 2017).

The probability of the absence of failures $P(t)$ is the probability that – under certain operating conditions – within the specified duration of operation, failure does not occur, and the probability of failure $Q(t)$ is the probability that – under the same conditions – a failure occurs during the specified time. The mill of serviceability (absence of failures) and malfunctions (presence of failures) of the system is incompatible and opposite events. The sum of the probabilities of such events, as is known from probability theory, is equal to unity. That is, the probability of failure and the probability of failure are related by:

$$P(t)+Q(t)=1 \tag{6.1}$$

as defined

$$\left.\begin{aligned} P(t) &= R(T \geq t); \\ Q(t) &= R(T \leq t); \end{aligned}\right\} \tag{6.2}$$

where

R is the probability symbol of an arbitrary event,
T is the operating time of the system to failure,
t is the operating time of the system for which we determine the reliability.

By the definition of probability theory (Sharov et al. 2015), the probability distribution function $F(x)$ of a random variable is the probability that the quantity will take a value less than some quantity x, that is,

$$F(x)=R(\xi<x) \tag{6.3}$$

It follows that the function of the probability of failure $Q(t)$ is similar to the distribution function of the operating time of the system to failure.

In a statistical assessment, the empirical probability of the absence of failures is defined as the relationship:

$$P_e(t)=\frac{N_0-n(t)}{N_B}=\frac{N_0}{N_0}-\frac{n(t)}{N_0}=1-\frac{n(t)}{N_0}, \tag{6.4}$$

and the empirical probability of failure as a relationship

$$Q_e(t)=n(t)/N_0 \tag{6.5}$$

where

N_0 is the number of nodes of the hydraulic pulse drive,
$n(t)$ is the number of hydroimpulse drive units that failed during time t.

The values of the empirical probabilities of the absence of failure and failures obtained by a statistical method always differ from theoretical ones (Chernovol et al. 2010). With an increase in the number of tested nodes $Pe(t)$ and $Qe(t)$, they asymptotically approach $P(t)$ and $Q(t)$. The same can be said about other quantitative characteristics of reliability. The initial conditions of the functions $P(t)$ and $Q(t)$ are defined in this way, at $t=0$ the hydro-pulse drive retains its original characteristics and meets the requirements presented to it, that is:

$$P(0)=1; \quad Q(0)=0. \tag{6.6}$$

Like any continuous function, the failure probability $Q(t)$ can be differentiated for all values of the argument. In probability theory, the derivative of the distribution function is called the distribution density:

$$f(x)=dF(x)/dx, \tag{6.7}$$

where

$f(x)$ is the probability density of a random variable ξ.

In the reliability theory, this density of the distribution of the system's operating time k for failures is called the failure rate $a(t)$ (Chernovol et al. 2010; Kukharchuk, Kazyv et al. 2017). We carry out the following transformations:

$$Q(t)=1-P(t). \tag{6.8}$$

Let us find the differential of the left and right sides of the dependence (6.8):

$$dQ(t)/dt=a(t)=\frac{d}{dt}[1-P(t)]=-dP(t)/dt. \tag{6.9}$$

Integrating the left and right sides of equality (6.9), we obtain:

$$Q(t)=\int_0^t a(t)\,dt \tag{6.10}$$

$$P(t)=1-\int_0^t a(t)\,dt. \tag{6.11}$$

By definition, the failure rate is the ratio of the number of nodes that failed per unit time to the number of all nodes that are tested, provided that they are not restored and are not replaced by serviceable ones

$$a_e(t)=n(\Delta t)/N_0\Delta t, \tag{6.12}$$

Here, $n(\Delta t)$ – the number of nodes that failed in the time interval Δt.

A typical time dependence of the failure rate is shown in Figure 6.1.

Three gaps are highlighted on the curve. Gap 1 is caused by a large number of failures at the beginning of operation of the hydro-pulse drive due to gross defects of its elements, errors of the operating personnel. The initial period is different for various hydroimpulse occasions. It can be reduced, or completely removed, using the methods of training and testing.

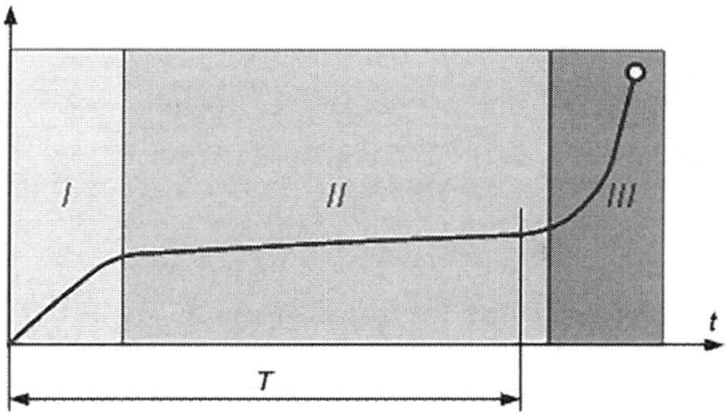

Figure 6.1 Typical failure rate a versus time *t*.

Gap 2 characterizes the normal operation of the hydraulic drive. The failures in this period are mostly unexpected in nature, their average frequency decreases.

The aging period is caused by the wear of the hydro-pulse drive, when, due to the aging of the elements (nodes), the failure rate gradually increases (Kukharchuk, Hraniak et al. 2016,; Kukharchuk, Bogachuk et al. 2017).

If the failure rate makes it possible to assess the reliability of the hydro-pulse drive for the desired period of time without taking into account the time of the previous operation, then the failure rate accounts for this effect. The distribution density, which takes into account the previous value of a random variable, is called the conditional density. Thus, the failure rate is the conditional density of the distribution of the failure time, which represents the instantaneous failure rate of the system at time *t*, provided that there are no failures up to this point.

The failure rate is defined as the ratio of the number of nodes of the hydraulic pulse drive that failed per unit time to the average number of nodes that worked correctly in a given period of time, provided that the nodes that failed to be restored and not replaced by serviceable ones

$$\lambda_e(t) = n(\Delta t)/N_{cep}\Delta t, \tag{6.13}$$

where

$N_{cep} = (N_i + N_{i+1})/2 = N_0 - n(\Delta t)$ is the average number of serviceable nodes at the beginning and end of the time interval Δt.

The probability representation of intensity was obtained using the main theorems of probability theory (Sharov et al. 2015). The proposed approach is new, which has its own elements of the novelty of the relation between the reliability theory and probability theory and mathematical statistics.

From expression (6.13), we replace $n(\Delta t)$ with the values obtained from formula (6.12), and N_{cep} – with its value from expression (6.4), we obtain:

$$\lambda(t) = a(t)/P(t) = -\frac{dp(t)}{dt}\Big/P(t). \tag{6.14}$$

In accordance with this, we finally define one more way $P(t), Q(t), a(t)$:

$$P(t) = \exp\left(-\int_0^t \lambda(t)\,dt\right), \tag{6.15}$$

$$Q(t) = (1 - \exp(-\lambda(t)dt)), \tag{6.16}$$

$$a(t) = \lambda(t)\exp\left(-\int \lambda(t)\,dt\right). \tag{6.17}$$

The obtained expressions (6.15), (6.16), and (6.17) establish the relationship between the probability of no failure, the probability of failure and the failure rate of the nodes of the hydraulic pulse drive. The result of this approach is the determination of mathematical expectation T, variance $D(t)$, and standard deviation $\sigma(t)$ as compound probability theory (Sharov et al. 2015). We will use the failure rate $\lambda(t)$, failure probability $P(t)$, failure probability $Q(t)$ as part of a reliability theory (Chernovol et al. 2010; Vedmitskyi et al. 2017).

Let us define the mathematical expectation T. From the point of view of probability theory, this is the mathematical expectation of the average value of a point estimate $\bar{t}_i$ of the average time, as the average operating time of the i-th node. Here, the indicators are reliability characteristics calculated using the tools of the mathematical apparatus of probability theory and mathematical statistics. Thus, the mean failure time T is the mathematical expectation of the operating time of the corresponding hydraulic pulse drive unit to failure.

In probability theory, the mathematical expectation of a random continuous variable ξ is called an integral of the form $\int xf(x)dx$. Turning to the theory of reliability, we can write:

$$T = \int_{-\infty}^{+\infty} ta(t)\,dt. \tag{6.18}$$

Substituting the value $a(t)$ with (6.17) in expression (6.18), integrating by parts and taking into account that $P(0) = 1$, $P(\infty) = 0$, and time cannot be negative, we obtain:

$$T = \int_{-\infty}^{+\infty} tP'(t)\,dt = -tP(t)\int_0^{\infty} + \int_0^{\infty} dt = \int_0^{\infty} P(t)\,dt. \tag{6.19}$$

Given the formula (6.17), we get:

$$T = \int_0^{\infty} \exp\left(-\int_0^{tt} \lambda(t)\,dt\right)dt \tag{6.20}$$

Expression (6.19) shows that the average time of absence of failures T is completely determined by the probability of the absence of failures $P(t)$ and represents the area that limits the curve $P(t)$ and the coordinate axes.

In order to determine the average time of absence of failures with statistical empirical data, we use the formula of a small sample of the form:

$$T_e = \sum_{i=1}^{N} t_i / N_0, \tag{6.21}$$

where

t_i is the operating time of the i-th hydropulse drive unit before a failure occurs.

This quantitative characteristic is important, as it allows in some cases to visually judge the reliability of hydraulic pulse drive units.

When assessing reliability using the average time of absence of failure, it is necessary to know the variance of the time of occurrence of failure $D(t)$, which characterizes the discrepancy of the studied value. We define it as the mathematical expectation of the squared deviation of a random variable t from the mathematical expectation of this random variable (T):

$$D(t)=\int_0^{\infty}(t-T)^2 a(t)\,dt. \tag{6.22}$$

Moreover, we note that it is necessary to minimize. We will develop this direction in further studies. For example, in classical sources, it is indicated that $D(t)=1$.

At the variance level $D(t)$, the root-mean-square deviation of the no-failure time is important. The standard deviation is:

$$\sigma(t)=\sqrt{D(t)}. \tag{6.23}$$

It is quite complete and simple to determine all quantitative characteristics of reliability from the law of distribution of the operating time of nodes to failure. Time is a random continuous quantity; therefore, arbitrary continuous distributions that are used in probability theory can be used as theoretical distribution laws.

6.3 CONCLUSIONS

The assessment of the efficiency and reliability of the operation of hydraulic pulse drives was conducted to evaluate the effectiveness and reliability of the introduced quantitative characteristics that are probabilistic in nature. They can be determined not only experimentally but also by theoretical analysis, where they are examined from the statistical and probabilistic points of view.

REFERENCES

Adamchuk, V.V., Bulhakov, V.M., Kaletnyk, H.N. & Kutsenko A.H. 2017. Yspolzovanye priamoho metoda hranychnikh elementov pri issledovanii statsyonarnikh kolebanyi plastin [The use of the direct method of boundary elements in the study of stationary oscillations of the plates]. *Vibratsii v tekhnitsi ta tekhnolohiiakh* 1(84): 8–14. [In Russian].

Chernovol, M.I., Chyrkun, V.I., Aulintainshi, V.V., Zazah, red., Chernovola, M.I. 2010. *Nadiinist silsko-hospodarskoi tekhniky. Reliability of Agricultural Machinery*. Kirovohrad: KOD, 320 p. [In Ukrainian].

Iskovych-Lototsky, R.D., Zelinska, O.V. & Ivanchuk, Y.V. 2018. *Tekhnolohiia modeliuvannia otsinky parametriv formoutvorennia zahotovok z poroshkovykh materialiv na vibropresovomu obladnanni z hidroimpulsnym pryvodom [Technology of Modeling of Estimation of Parameters of Forming of Billets from Powder Materials on the Vibropress Equipment with the Hydropulse Drive*]. Vinnytsia: VNTU, p. 152. [In Ukrainian].

Iskovych-Lototsky, R.D., Zelinska, O.V., Ivanchuk, Y.V. & Veselovska N.R. 2017. Development of the evaluation model of technological parameters of shaping work pieces from powder materials. *Eastern-European Journal of Enterprise Technologies* 1(85): 9–17

Ivanov, M.I., Pereiaslavskyi, O.M., Sharhorodskyi, S.A. & Kovalova, I.M. 2001. Parametrychne zbudzhennia pulsatsii pidchas roboty rehulovanoho aksialnoho rotornoporshnevoho nasosa [Parametric excitation of pulsations during operation of an adjust ableaxial rotary piston-pump].Cherkasy, pp. 151–152. MaterialyXXII mizhnarodno I naukovo-tekhnichnoi konferentsii" Hidroaeromekhanika v inzhenernii praktytsi" (m. Cherkasy 23–26 trav. 2017 r.) [In Ukrainian].

Kukharchuk, V.V., Bogachuk, V.V., Hraniak, V.F., Wójcik, W., Suleimenov, B. & Karnakova, G. 2017. Method of magneto-elastic control of mechanic rigidity in assemblies of hydropower units. *Proceedings of the SPIE* 10445: 104456A.

Kukharchuk, V.V., Hraniak, V.F., Vedmitskyi, Y.G., Bogachuk, V.V., Zyska, T., Komada, P. & Sadikova, G. 2016. Noncontact method of temperature measurement based on the phenomenon of the luminophor temperature decreasing. *Proceedings of the SPIE* 10031: 100312F.

Kukharchuk, V.V., Kazyv, S.S. & Bykovsky, S.A. 2017. Discrete wavelet transformation in spectral analysis of vibration processes at hydropower units. *Przeglad Elektrotechniczny* 93(5): 65–68.

Rutkevych, V.S. 2017. Adaptyvnyi hidravlichnyi pryvod blochno-portsiinoho vidokremliu vachakonservovanoho kormu [Adaptive hydraulic actuator of block-portion separator of canned food]. *Tekhnika* 4(99): 108–133, enerhetyka, transport APK [In Ukrainian].

Sharov, S.V., Lubko, D.V. & Osadchyi, V.V. 2015. *Intelektualni informatsiini systemy: navch. posib.* [*Intelligent Information Systems: A Textbook*], Melitopol: Vyd-vo MDPU im. B. Khmelnytskoho, 144 p. [In Ukrainian].

Vedmitskyi, Y.G., Kukharchuk, V.V. & Hraniak, V.F. 2017. New non-system physical quantities for vibration monitoring of transient processes at hydropower facilities, integral vibratory accelerations. *Przeglad Elektrotechniczny* 93(3): 69–72.

Zhu, Z.Q., Lee, B., Huang, L. & Chu, W. 2017. Contribution of current harmonics to average torque and torque ripple in switched reluctance machines. *IEEE Transactions on Magnetics* 53(3): 1–9.

Chapter 7

The Use of Cyanobacteria – Water Pollutants in Various Multiproduction

Mykhaylo V. Zagirnyak and Volodymyr V. Nykyforov
Kremenchuk Mykhailo Ostrohradskyi National University

Myroslav S. Malovanyy, Ivan S. Tymchuk, and Christina M. Soloviy
Lviv Polytechnic National University

Volodymyr V. Bogachuk
Vinnytsia National Technical University

Paweł Komada
Lublin University of Technology

Ainur Kozbakova
Institute of Information and Computational Technologies CS MES RK
Almaty University of Power Engineering and Telecommunications

Zhazira Amirgaliyeva
Institute of Information and Computational Technologies CS MES RK
Al-Farabi Kazakh National University

CONTENTS

7.1 INTRODUCTION

The financial instability of the world economy because of an energy crisis brings into focus the search for new nonconventional (alternative) energy sources. Among others, these include solar energy, accumulated in the biomass of photosynthetic (autotrophic) plants (so-called solar energy bioconservation). It should be noted that to date, a

DOI: 10.1201/9781003177593-7

certain portion of the energy potential of land-based plant biomass is already utilized by the humankind. A sixth of the energy consumed is produced from agricultural and other phytomass. This is equivalent to the daily use of more than 4 million tons of oil. However, the biomass of aquatic microorganisms and phytoplankton (microalgae and cyanobacteria) is not in demand at all.

In recent years, the European water reservoirs have been subject to strong anthropogenic pressure leading to their eutrophication and overgrowing processes. Significant environmental problems are associated with the pollution of water bodies not fully treated from municipal wastewater (Sakalova et al. 2019a), landfill filtrates (Malovanyy et al. 2019a), and heavy metals (Sakalova et al. 2019b). The mechanisms of eutrophication and the dynamics of the ecological factors associated with it (overgrowth of vascular aquatic plants, pH, BOD_5, the concentration of nutrients, chlorophyll "a" and oxygen) were studied under the conditions of bio-hydrocenoses of different types of water bodies in Poland (Sojka et al. 2019), Romania (Dughilă et al. 2012), Serbia (Grabić et al. 2016), and other countries of Eastern Europe.

The Dnieper River is one of the largest waterways in Europe, which provides fresh water for 65% of agricultural and 75% of industrial products, as well as 80% of the Ukrainian population. The construction of a cascade of six reservoirs in the mid-20th century not only increased the Dnieper economic potential but also confronted Ukraine with a number of new environmental problems caused by the water blooming. This seasonal (from June to September) phenomenon is the massive development of toxic blue-green planktonic microalgae (cyanobacteria), due to the large-scale eutrophication of low-flowing water bodies. Previously, the water reservoirs of the Dnieper cascade (Pinchura et al. 2018) were classified using the Trophic State Index (TSI), which was developed by the Florida Department of Environmental Protection. The dynamics of TSI for the period from 1986 to 2016, as well as the values of the concentration of phosphorus and chlorophyll "a", were discussed too. Its presence is characteristic of the species of all the real algae divisions (Phycobionta) and vascular plants (Embryobionta).

The question of the spectrophotometric determination of bacterio-chlorophyll, the dominant pigment of the photosynthetic system of blue-green (not real) algae, which are prokaryotic unicellular organisms with palmeloid and trichal type thallus, is still an open question.

The water blooming, the dominant agent of which in the Dnieper reservoirs conditions is the cyanobacterium *Microcystis aeruginosa* Kützing (Figure 7.1), should be regarded as an environmental signal of the structural and functional organization violation of aquatic ecosystems (biohydrocenosis) as a result of the acute increase of the nitrogen, phosphorus, and carbon content available for autotrophic hydrobionts. This problem, together with the increased prices for natural gas, exported by Ukraine, prompted us the idea of using the cyanobacteria biomass for biogas production (Shmandii et al. 2010).

Earlier studies (Nykyforov et al. 2016) found that 1 kg of dry microcystis biomass contains 320 g of carbohydrate, 300 g of protein, 5 g of fat, 3.8 mg of thiamine, 3.7 mg of nicotinic acid, 0.9 mg of pyridoxine, and 0.1 mg of biotin. The hydrolysis of proteins and polysaccharides produced 17 aminoacids and 11 monosaccharides, respectively. It is also known that the maximum bloom of natural population of *microcystis*

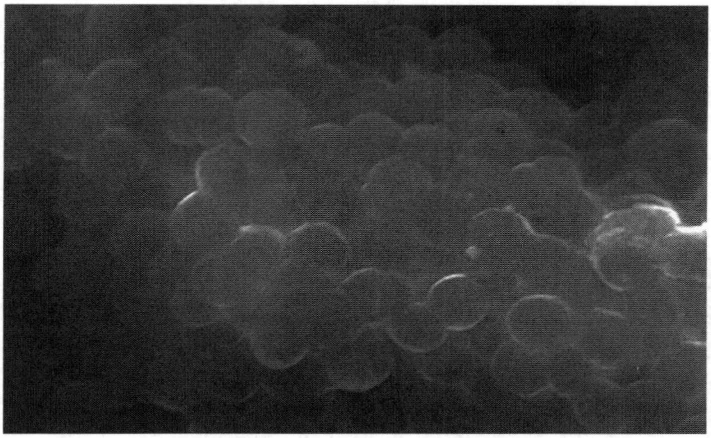

Figure 7.1 SEM image of Microcystis aeruginosa Kützing colony, 3,000$^{\times}$.

aeruginosa (Kützing) Kützing of the Kremenchuk Reservoir (Ukraine) is accompanied by the highest content of carbohydrates and vitamins in the biomass.

Thus, the blue-green algae are a source of biologically active substances that can be used not only in agriculture for feed vitaminization but also for the biotechnological and nanoprocesses. Herewith, the biologically active substances of BGA in microquantities can be purposefully delivered to specific cells of tissues and organs using magnetic particles – the carriers of biologically active substances – for the desired therapeutic effect. However, this unit requires additional targeted research.

Removing their excess amount from the reservoirs will not only improve the natural water quality and restore the chemical and biological regimes of these ecosystems but will also enable to obtain cheap biofuel and free fertilizer for farms (Table 7.1).

With a minimum of primary production (50 kg/m^3) of littoral (depth up to 2 m; 18.5% of the water area) only at the Kremenchug Reservoir (2,250 km^2), about $4.14 \cdot 10^7$ tons of biomass can be collected annually during the vegetative period (70–120 days), which is about 2.51×10^5 tons of dry organic matter. Owing to biomethanogenesis, it is possible to obtain up to 18.84 million m^3 of methane or to extract about 11,000 tons of

Table 7.1 Environment and economic problems and the results of new biotechnologies implementation

Problems	*Results*
Impaired quality of drinking water	Free renewable energy resource (cyanobacterial biomass)
Massive suffocation of fish and other hydrobionts	Production of cheap biofuel and organic-mineral fertilizers
Reduced recreation in the water area and coastal territories	Improving drinking water quality
High price for exported gas	Littoral ecosystems rehabilitation

lipids for the biodiesel production, as well as to obtain up to 25 million tons of liquid biological fertilizers (Malovanyy et al. 2016).

7.2 MATERIALS AND METHODS

The REM-106 scanning electron microscope was used for the research of cyanobacterial cells. Its main characteristics include the maximum image size (1280 × 960 px), pressure adjustment range in the chamber (1–270 Pa), accelerating pressure range (0.5–30.0 kW), and magnification adjustment range ($15–3 \times 10^5$).

The measurement of the mass fraction (%) of the basic chemical elements in the cyanobacterial dry biomass was performed using the EXPERT-3L X-ray fluorescence analyzer. The basic parameters are the range of the chemical elements: from ^{12}Mg to ^{92}U, the range of mass fractions (concentrations) of elements: 0.01%–99.90% and an equivalent dose rate of X-ray radiation on the instrument surface does not exceed 74 nSv/h.

The toxic effect of the water solutions of cyanobacterial substrate and digestate was determined in accordance with the Water Framework Directive (2000/60/EC) and International Organization for Standardization: Water Quality – Determination of the Inhibition of the Mobility of Daphnia magna Straus (Cladocera, Crustacea) – Acute Toxicity Test (ISO 6341:1996).

In laboratory studies, an LTI–500 optical quantum laser generator was used: the wavelength of the beam was 1,062 nm, and the pulse frequency was 50–1,000 Hz (Figure 7.2). In the experiment, laser processing of the cyanobacteria biomass continued until it was heated to a temperature of +46°C, which triggers the mechanism of photothermal destruction of the surface structures of bacterial cells with a release of intracellular biomolecules.

In order to perform acoustic cavitation, a suspension of cyanobacteria was introduced into an ultrasonic reactor (Figure 7.3). Ultrasonic oscillations (frequency – 22 kHz, power – 35 W, intensity – 1.65 W/cm^3/unit volume) from the UZDN-2T generator were

Figure 7.2 Laser irradiance facility.

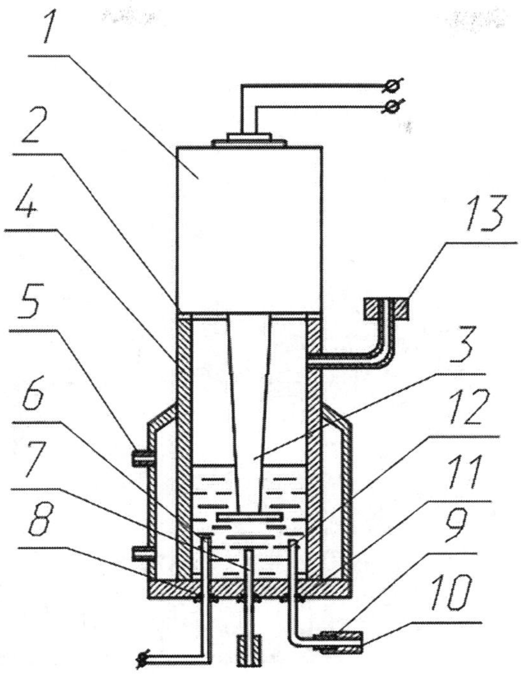

Figure 7.3 Reactor scheme for the processing cyanobacteria suspension by ultrasound: 1 – magnetostrictor, 2, 8, 9 – seals, 3 – waveguide, 4 – reactor, 5 – sleeve for coolants, 6 – thermocouple, 7 – sleeves for gas input, 10, 11 – cap nuts, 12 – sampler, 13 – gas outlet sleeves.

transmitted using a magnetostrictive emitter immersed in the test medium ($V = 150\,cm^3$). Throughout the process, CO_2 was bubbled through the test suspension. The reactor was continuously cooled under running water. The conditions for ultrasonic treatment are as follows: $T = 298$ K, $P = 1 \times 10^5$ Pa, $\nu_{UZ} = 22$ kHz (Malovanyy et al. 2019b).

Determination of the treatment efficiency of cyanobacterial suspension in the field of hydrodynamic cavitation was performed in a hydrodynamic cavitator (Figure 7.4), where a three-bladed wedge-shaped impeller with a sharp front and blunt trailing edges was used as a cavitating body, the impeller speed was 4,000 rpm. In the working tank of the cavitator was poured 1 liter of suspension of cyanobacteria (Nykyforov et al. 2019, Kukharchuk et al. 2017, Vedmitskyi et al. 2018).

In order to establish the prospects of vibrocavitation treatment for water purification and water preparation of public water bodies, the current model of vibrocavitator was used (Figure 7.5), the main components of which are a cylindrical working chamber with a volume of 1 dm^3, blue-green algae substrate and gases supply system, electromagnetic vibratory drive with attached oscillating deck with cavitation perturbators and electric power supply network of the vibrating drive. The study on the influence of vibrocavitation treatment on the parameters of the model mixtures selected for experiments was performed at a frequency of 110 Hz in a nitrogen medium (Petruk et al. 2015, Vasilevskyi et al. 2017).

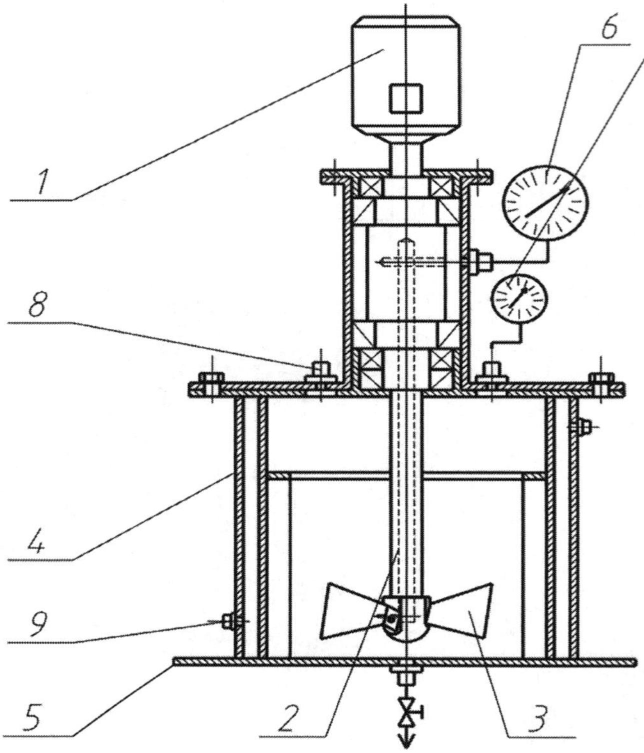

Figure 7.4 Schematic representation of the laboratory-experimental stand: 1 – electric motor; 2 – shaft; 3 – cavitation body; 4 – working volume; 5 – stand (frame); 6 – tachometer; 7 – manometer; 8 – samplers; 9 – sleeve for coolant supply.

7.3 RESULTS AND DISCUSSION

The results of the biomass quantitative analysis are indicating its high-energy potential (Table 7.2). One liter of cyanobacterial cell suspension contains 1.43 g of humic substances, 380 mg of lower fatty acids, and 205 mg of other organic compounds, which are the main substrates for methanobacteria. Because of their bioconversion, biogas is obtained with a high calorific value (18–25 MJ/m^3) due to significant methane content in it (75%–80%).

The high concentration of fulvic and humic acids in the digestate (965 mg/dm^3), as well as a balanced amount of the main biogenic elements (NPK – 12.7, 1.3 and 2.1%, respectively), allows considering it as a valuable organic-mineral fertilizer for culture plants (Tables 7.2 and 7.3).

The biotesting results of substrate aqueous suspensions and digestate using the lower crustaceans of *Daphnia magna* Straus indicate a toxicity decrease of the substrate as a result of methanogenesis by 7.5–12.7 times, because the main cyanobacterial toxin (microcystin) is also subjected to bioconversion during this process (Figure 7.6).

The limiting factor in the use of cyanobacteria as raw materials for different types of production is the diffusion resistance of cyanobacteria walls. Therefore, it is promising for

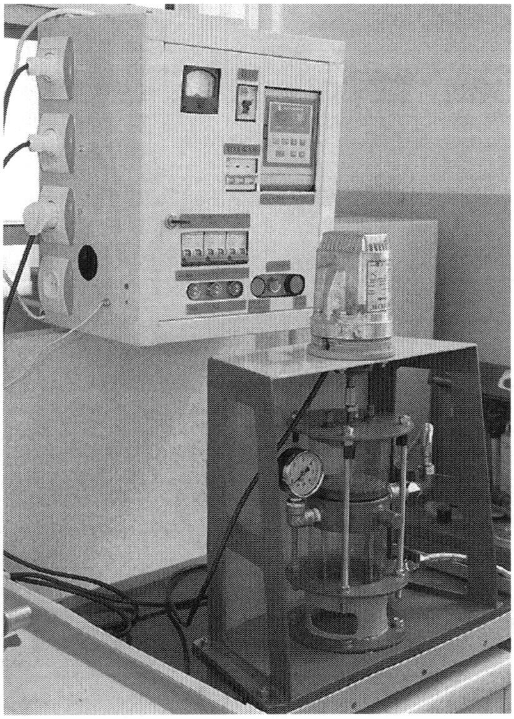

Figure 7.5 Active model of vibrocavitation.

Table 7.2 The content of acids (A) and other organic substances (B) in the cyanobacteria biomass before (substrate) and after (digestate) biomethanogenesis.

A	*Substrate*	*Digestate*	*B*	*Substrate*	*Digestate*
Organic content, mg/dm³					
Acetic	84	<20	Lactose	21	<20
Propionic	90	<20	Xylose	<20	<20
Oilyc	<20	<20	Glycerol	36	<20
Isobutyric	204	43	Acetyl glucosamine	30	<20
Valerianic	<20	<20	Gluconate racemate	98	91
Izovalerianic	<20	<20	Σ humus	1,425	965
Galacturonic	25	<20	Σ fatty acids	380	43
Fulvic	745	365	COD, g/kg	10.43	9.43
Humic	680	600	BOD_5, mg O/dm^3	1,750	530

Table7.3 Mass fraction of elements in the cyanobacteria biomass after methanogenesis

Element	^{12}C	^{14}N	^{16}O	^{24}Mg	^{28}Si	^{31}P	^{32}S	^{35}Cl	^{39}K	^{40}Ca	^{64}Cu
%	44.66	12.67	31.81	1.30	0.61	1.34	0.69	0.22	2.14	1.92	0.60

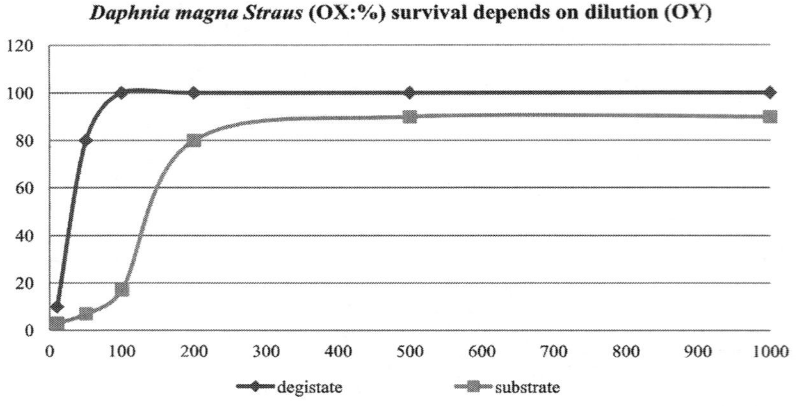

Figure 7.6 Biotesting results of substrate and digestate water solutions.

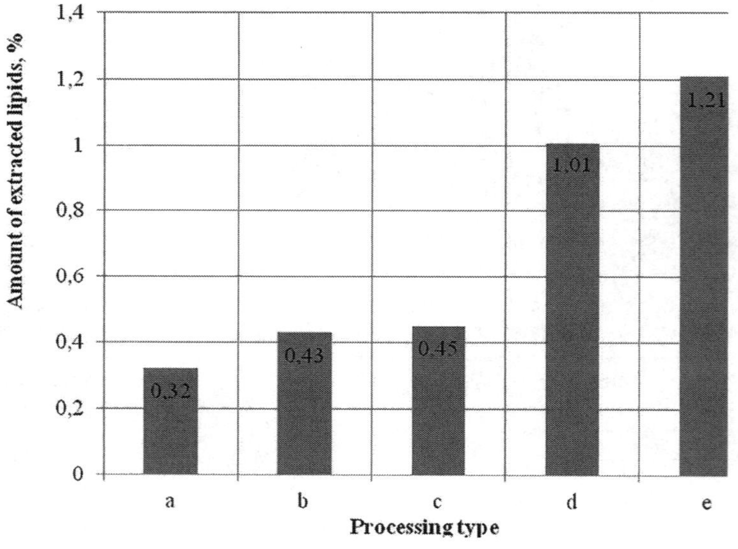

Figure 7.7 Dependence of the amount of lipids extracted from cyanobacteria (% of dry weight) on the type of pre-treatment of biomass: (a) without pre-treatment; (b) after laser treatment; (c) after processing in an acoustic cavitation field; (d) after processing in the field of hydrodynamic cavitation; and (e) after vibro-cavitation treatment.

all biotechnological processes to carry out the preliminary preparation of the cyanobacteria biomass, which would result in the destruction of the walls of cyanobacteria for more full discharge of their intracellular contents. The following methods of this preparation were considered: ultrasonic cavitation, hydrodynamic cavitation, and vibrocavitation. The number of lipids extracted from biomass served as a criterion for evaluating the efficiency of biomass pretreatment. The research results are shown in Figure 7.7.

As can be seen from Figure 7.7, laser treatment and acoustic cavitation field treatment are inefficient. Processing in the field of hydrodynamic cavitation can be promising for practical use, but the vibrocavitation treatment is the most promising. The technological advantage of such treatment may be the ability to implement processing biomass in a continuous stream mode.

7.4 CONCLUSIONS AND PERSPECTIVES

Thus, at present, we have tested obtaining three target products (biogas and/or bioalcohols production, obtaining organic mineral fertilizer and lipid extraction for biodiesel production) of environmental biotechnology for the cyanobacterial processing that are massively developing in the Dnieper reservoirs.

It has been established that pre-treatment of biomass, which allows destroying the walls of cyanobacteria and releasing intracellular substance, can be a promising way to increase the efficiency of biotechnological processes using cyanobacteria as a raw material. The most promising way of such pre-treatment is the use of the vibrocavitation field. The modern bio-based economy involves the development of a network of bio-clusters, the main structural unit of which is biorefinery – biomass conversion enterprises full cycle fuel, energy and chemicals. In this case, a biogas plant can be considered as a biorefinery that will become an energy center of the bio-cluster infrastructure (gas generator, biogas liquefaction plant, biomass fuel boiler, cogeneration unit, phycocyanin production plant, aminoacids hydrolyzate, processing of digestate for biofertilizer, greenhouse complex, etc.).

Currently, a SWOT analysis has been carried out and a business plan has been drawn up for the construction of a bio-cluster in the Middle Dnieper region for the production of fertilizers, biofuels of different generations, as well as the biologically active substances from the biomass of blue-green blue green algae and other mass forms of hydrobionts.

The research using standardized methods of light and electron microscopy revealed that the main type of blue-green algae of the Kremenchuk and Kam'yansk reservoirs of the Dnieper cascade (Ukraine) is blue-green microalgae weighing 4.14×10^7 tons during the growing season. Owing to biomethanogenesis, it is possible to obtain up to 18.84 million m^3 of methane or to extract about 11,000 tons of lipids for the biodiesel production, as well as to obtain up to 25 million tons of liquid biological fertilizers.

It was proven that algae residues can be used as biofertilizers in agriculture and forestry as a substrate water solution at various dilutions, the survival rate of cladocerans as test objects is from 80% to 100%. The optimum concentration of the spent substrate aqueous solution depends on the methanogenesis process completeness. The data of the X-ray diffraction analysis confirmed the compliance of the elemental composition of the spent substrate with the green mass of plants, which will provide a high nutritional value of algae residues.

ACKNOWLEDGEMENTS

This project was supported by the bilateral Austrian-Ukrainian Government Scientific Grant "Methods of the processing of cyanobacteria (algae) biomass, causing water reservoir blooming" (Grant No. SR 0117U003299, 2017–2018).

REFERENCES

Dughilă, A., Iancu, O.G. & Râşcanu, I.D. 2012. Geochemical evaluation of quality indicators for the water of the Tansa Lake from the Jijia catchment, Romania. *Carpathian Journal of Earth and Environmental Sciences* 7(3): 79–88.

Grabić, J., Benka, P., Bezdan, A., Josimov-Dunderski, J. & Salvai, A., 2016. Water quality management for preserving fish populations within hydro-system danube-tisa-danube, Serbia. *Carpathian Journal of Earth and Environmental Sciences* 11(1): 235–243.

Kukharchuk, V.V., Bogachuk, V.V., Hraniak, V.F., Wójcik, W., Suleimenov, B. & Karnakova, G. 2017. Method of magneto-elastic control of mechanic rigidity in assemblies of hydropower units. *Proceedings of the SPIE* 10445: 104456A.

Malovanyy, M., Moroz, O., Hnatush, S., Maslovska, O., Zhuk, V., Petrushka, I., Nykyforov, V. & Sereda, A. 2019a. Perspective technologies of the treatment of the wastewaters with high content of organic pollutants and ammoniacal nitrogen. *Journal of Ecological Engineering* 20(2): 8–15.

Malovanyy, M., Nykyforov, V., Kharlamova, O. & Synelnikov, O. 2016. Production of renewable energy resources via complex treatment of cyanobacteria biomass. *Chemistry & Chemical Technology* 10(2): 251–254.

Malovanyy, M., Zhuk,V., Nykyforov, V., Bordun, I., Balandiukh, I. & Leskiv, G. 2019b. Experimental investigation of *Microcystis aeruginosa* cyanobacteria thickening to obtain a biomass for the energy production. *Journal of Water and Land Development* 43(1): 113–119.

Nykyforov, V., Malovanyy, M., Aftanaziv, I., Shevchuk, L. & Strutynska, L. 2019. Developing a technology for treating blue-green algae biomass using vibro-resonance cavitators. *Naukovyi Visnyk Natsionalnoho Hirnychoho Universytetu* 6:181–188.

Nykyforov, V., Malovanyy, M., Kozlovs'ka, T., Novokhatko, O. & Digtiar, S. 2016. The biotechnological ways of blue-green algae complex processing. *Eastern-European Journal of Enterprise Technologies* 5/10(83): 11–18.

Petruk, V., Kvaternyuk, S., Kozachuk, A., et al. 2015. Multispectral televisional measuring control of the ecological state of waterbodies on the characteristics macrophytes. *Proceedings of the SPIE* 9816: 98161Q.

Pinchura, V., Malchykova, D., Ukrainskij, P., Shakhman, I. & Bystriantseva, A. 2018. Anthropogenic transformation of hydro-logical regime of the Dnieper River. *Indian Journal of Ecology* 45(3): 445–453.

Sakalova, H., Malovanyy, M., Vasylinych, T. & Kryklyvyi, R. 2019a. The research of ammonium concentrations in city stocks and further sedimentation of ion-exchange concentrate. *Journal of Ecological Engineering* 20(1): 158–164.

Sakalova, H., Malovanyy, M., Vasylinycz, T., Palamarchuk, O. & Semchuk, J. 2019b. Treatment of effluents from ions of heavy metals as display of environmentally responsible activity of modern businessman. *Journal of Ecological Engineering* 20(4): 167–176.

Shmandii, V., Nykyforov, V., Alferov, V., Kharlamova, E. & Pronin, V. 2010. Use of blue-green algae for biogas production. *Gigiena i sanitariia* 6: 35–37.

Sojka, M., Jaskuła, J., Wrózyński, R. & Waligórski, B. 2019. Application of sentinel-2 Satellite imagery to assessment of spatio-temporal changes in the reservoir overgrowth process – A case study: Przebędowo, West Poland. *Carpathian Journal of Earth and Environmental Sciences* 14(1): 39–50.

Vasilevskyi, O.M., Kulakov, P.I., Dudatiev, I.A., et al. 2017. Vibration diagnostic system for evaluation of state interconnected electrical motors mechanical parameters. *Proceedings of the SPIE* 10445: 104456C.

Vedmitskyi, Y.G., Kukharchuk, V.V., Hraniak, V.F., Vishtak, I.V., Kacejko, P. & Abenov, A. 2018. Newton binomial in the generalized Cauchy problem as exemplified by electrical systems. *Proceedings of the SPIE* 10808: 108082M.

Chapter 8

Elaboration of Biotechnology Processing of Hydrobionts Mass Forms

Sergii V. Digtiar and Volodymyr V. Nykyforov
Kremenchuk Mykhailo Ostrohradskyi National University

Mykhailo O. Yelizarov
Kremenchuk Mykhailo Ostrohradskyi National University
Vinnytsia National Technical University

Myroslav S. Malovanyy
National University Lviv Polytechnik

Tatyana N. Nikitchuk
Zhytomyr Polytechnic State University

Andrzej Kotyra
Lublin University of Technology

Saule Smailova
East Kazakhstan State Technical University named after D. Serikbayev

Aigul Iskakova
Satbayev University

CONTENTS

8.1 INTRODUCTION

The explosive formation of the blue-green algae biomass (cyanide or cyanobacteria) in the cascade of Dnieper reservoirs is already seasonal, causing an imbalance in hydroecosystems. Development and implementation of the technological process of methane production and fertilizers from their biomass is one of the promising natural

DOI: 10.1201/9781003177593-8

biotechnologies that can reduce the severity of the "blooming" problem of reservoirs, minimize the environmental risks, and provide the region with additional energy resources.

The search for effective technologies for biofuel production is conducted within the framework of the program of diversification of natural gas sources in Ukraine. Solving the problems with energy supply of small farms with cheap biofuels based on the organic matter extracted from alternative sources will allow optimizing technological processes depending on the season and available raw materials (Nykyforov et al. 2008, 2011, Buzovsky et al. 2008).

The purpose of this chapter was to develop a new technological process for the production of biogas and organo-mineral fertilizers based on the processes of biodegradation and bioconversion of organic matter of aquatic organisms, consisting mainly of cyanobacteria.

In order to achieve this goal, the following tasks were set and solved:

- the species composition of the initial substrate for biomethanogenesis and its microbiological characteristics, as well as the ecological and economic significance of cyanobacteria and prospects for efficient use of their biomass were determined;
- physicochemical and biological aspects of the process of biomethanogenesis and features of biochemical processes of biogas production were studied, the process of biomethanogenesis was modeled under laboratory conditions, the chemical composition of the biogas samples of different origin was determined and the comparative analysis of their physical properties was carried out;
- a new waste-free technological process that ensures the rational use of natural resources (excess biomass of aquatic organisms) and diversifies the production of second-generation biofuels was created;
- the scientific principles of safe technology of cyanide biomass processing on the basis of the biomethanogenesis process were substantiated;
- the project of biomethanogenic installation was executed and technical conditions of process of processing of mass forms of aquatic organisms with obtaining of methane and organo-mineral fertilizer were developed;
- the use of different substrates for the production of a biogas methane-containing mixture was studied; technical conditions for the target products of biotechnology were developed;
- a mathematical model of the biomethanogenesis process with the method of central composite rotatable planning of the complete factorial experiment CFE-2 with star points was developed; and a virtual complex for control and automation of the technological process of methane and fertilizer production from mono- and multisubstrates was created;
- technical and economic efficiency of bioconversion of organic mass of cyanobacteria was determined.

8.2 MATERIAL AND RESEARCH RESULTS

The analysis of the literature data revealed the scale and periodicity of the "blooming" process, which in turn convincingly proves the prospects of using the biomass of aquatic organisms as a substrate for biomethanogenesis, taking into account the regional conditions of local hydroecosystems. The species composition of phytoplankton was determined, in particular, the dominant role of the cyanobacteria of the *Microcystis aeruginosa* Kützing species in the "blooming" spots was indicated (Figure 8.1).

This species belongs to the order of Chroococcales. Although morphologically (the absence of a formed nucleus and many membrane organelles in the cell, the bacterial type of photopigments, etc.), this species belongs to bacterial organisms and according to the classification of A. L. Takhtadzhyan belongs to class Chroococcophyceae division Cyanobacteria subregnum Oxiphotobacteriobionta. In the scientific literature, it is traditionally mentioned under the common name "blue-green algae" (Buzovsky et al. 2008, Barsukova et al. 2005, Lemeza 2008).

It was proven that when the ecological balance in hydrobiocenoses is disturbed under the influence of anthropogenic factors, the processes of autoregulation of their formation, which are most balanced in balanced terrestrial phytocenoses, change significantly. This affects the level of accumulation of biologically active substances in the aquatic environment, which affects the formation of water quality and functional activity of aquatic organisms.

During the mass extinction of blue-green algae due to the accumulation of decomposition products, which are easily oxidized, the concentration of oxygen in the water decreases until complete disappearance. As a result, there are conditions for freezing and mass death of fish (July–August 2001–2004 and 2013–2016) (Nykyforov & Kozlovs'ka 2002, Nykyforov & Avramenko 2014). Oxygen deficiency, which was previously observed in the river very rarely, after the regulation of runoff becomes huge in a large area of the Dnieper cascade reservoirs. The alkalinity of water during the "blooming" in the upper layer increases to 9.6. In summer, carbon dioxide is usually absent in the surface layers, at the same time, in the bottom layers of water at a depth of 15–20 m, its concentration reaches 19.0 mg/dm^3. The content of nutrients in the water of the reservoir varies in different seasons of the year in a wide range (Grebin et al. 2014).

The high degree of accumulation of the nitrogen and phosphorus compounds in the water of the reservoir significantly affects the intensive development of planktonic

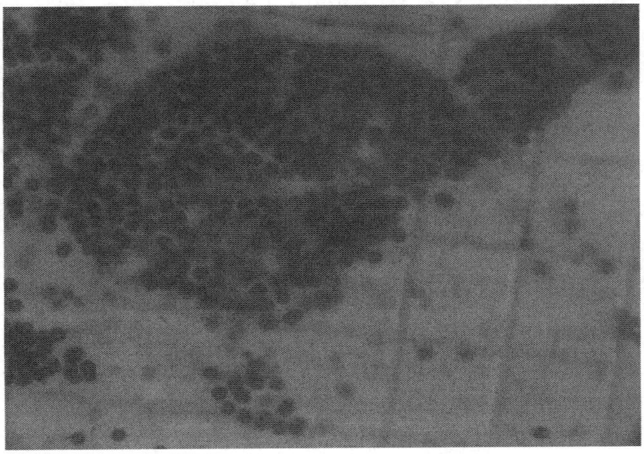

Figure 8.1 Microcystis aeruginosa.

blue-green algae. After the extinction of these algae on the surface of the reservoir, huge accumulations of nutrients and bacterial agglomerations with a characteristic smell of skatole are formed. The latter is a decomposition product of tryptophan contained in the biomass of blue-green algae, and additionally formed indole with a specific odor and serine, which under the action of the sulfate-reducing enzymes decomposes to the corresponding tolyl ethers with an unpleasant odor of methyl mercaptan, which gives natural waters odor. The data on the substrate and biological agents of the developed technology were also given.

The main object of research is ecological (environmental) biotechnology of processing the excess biomass of BGA and other aquatic organisms, which is formed in the course of the eutrophication of water during its “blooming”.

The subject of our research is cyanobacteria, the biomass of which should be used as a new restorative substrate for useful products. Different groups of prokaryotes act as a substrate and bioagents. In the first case, these are mainly the representatives of photoautotrophic microorganisms – producers of primary biomass in trophic chains. In the second, a complex of chemoautotrophic bacteria-symbionts that ensure the decay and destruction of this biomass to obtain the final products that can be effectively used by humans to meet their economic needs. The following empirical methods were used during the research:

- microscopic methods using Ningbo Shengheng XS-3330 optical microscope and PEM–106 B electron microscope (Gregirchak 2009, Perekrestov 2014);
- laboratory-analytical methods to determine the physicochemical properties of the biogas samples, X-ray fluorescence analysis using the EXPERT 3L analyzer to measure the mass fraction (%) of basic chemical elements in the samples of substrate biomaterial, absorption analysis to determine the components in the biogas samples using Кристал – 2000M gas chromatograph (methods of analysis according to the State Standarts of Ukraine (DSTU) ISO 6974-1:2007, DSTU ISO 6974-4:2007);
- biotesting methods for determining the phytotoxic effect of organo-mineral fertilizers on crops and for detecting the acute toxic effect of the substrate spent during bioconversion on living test objects (methods of analysis in accordance with the National Standard of Ukraine: Water quality. ISO 6341:1996, MOD according to the State Standarts of Ukraine (DSTU) 4173:2003;
- mathematical method of variation statistics for processing the obtained experimental data and assessing the degree of their reliability (Kustovska 2005);
- method of system analysis to obtain and summarize the results and calculations, as well as the method of experiment planning (Nykyforov et al. 2016a).

The research program, which reflects their main directions, and the relationship between the stages of solving the tasks are presented in Figure 8.2.

During microscopy, the average diameter of microcystis cells was determined to be 3.14 μm (Digtiar 2016a). The volume of its average spherical cell is about 15 μm^3 and its mass is 15 μg. The results of microscopy of the samples showed that the number of cells in them can reach 1 million/cm^3 and more. Thus, in the Kremenchuk and Kamyansk reservoirs, the mass of *Microcystis aeruginosa* Kützing is 4.14×10^7 tons during the growing season (Pastorek et al. 2006), which in terms of dry weight can range from 1.61 to 2.50×10^5 tons of organic matter.

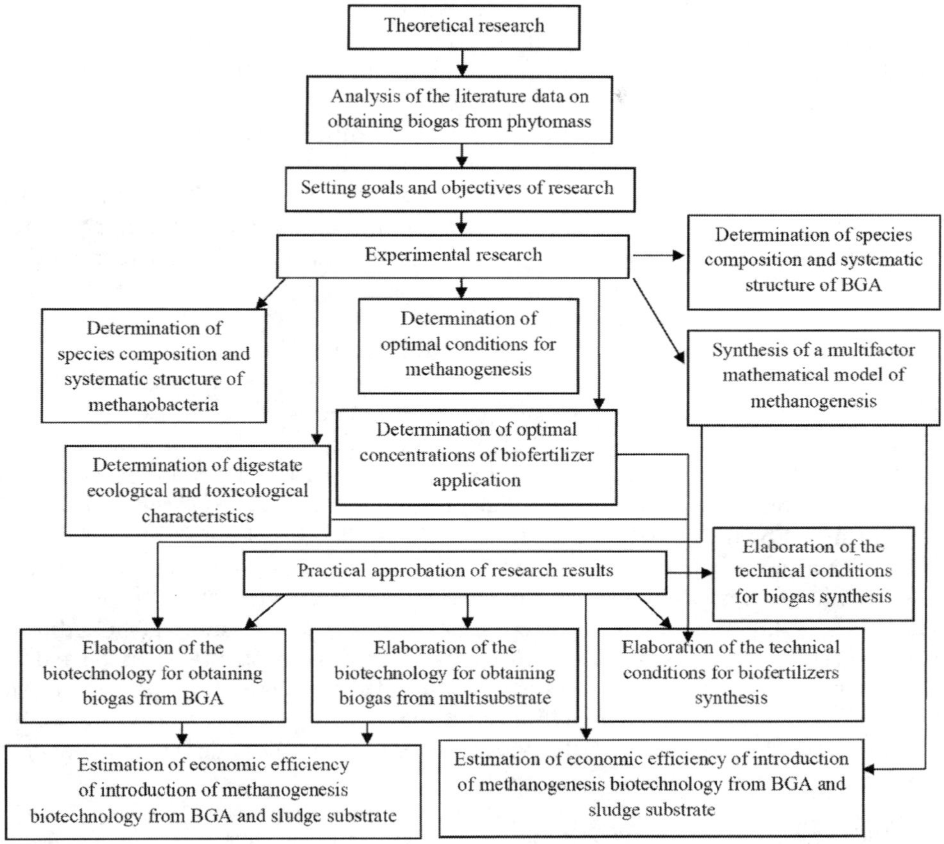

Figure 8.2 Research program.

The review of other substrates that are traditionally used to obtain a biogas mixture: agricultural waste (both livestock and crop), as well as activated sludge in water treatment plants – a mixture of aerobic microorganisms that can sorb and oxidize pollutants in wastewater - was performed. The quality of activated sludge depends on the type and amount of organic contaminants, the presence of toxic impurities, the completeness of the preliminary settling, the duration and intensity of aeration, the load on the activated sludge and so on.

It was proven that the degradation of organic compounds during biomethanogenesis is carried out as a multistage process in which their molecular carboskeleton is gradually destroyed by numerous groups of microorganisms (Nykyforov et al. 2018, Kvaternyuk et al. 2018a). The stages of methanogenesis, their sequence and features, as well as the groups of microorganisms that ensure the course of each of them were considered. The biochemical issues of destruction of organic matter of cyanobacteria and its transformation into other, more convenient for economic consumption, macroergic compounds were revealed. An attempt to investigate the microbiological aspects of biomethanogenesis was made, in particular to determine the optimal characteristics of the environment that ensure maximum efficiency of the process.

The influence of substrate quality on the process of biogas formation was clarified. On the one hand, the diversity of the composition of the methanogenic association of bacteria allows processing almost any organic waste. On the other hand, in order to maintain the required level of vital activity of all microorganisms, it is necessary to observe the ratio of the content of Nitrogen to Carbon12 (according to some sources up to 30) to one in the substrate. If the ratio of C:N is excessive (>>30), then the lack of nitrogen causes a slowing down of the methane fermentation process, if it is too small (<<10), then a significant amount of ammonia is formed, which is toxic to bacteria. The content of organic compounds in fresh biomass of blue-green algae and digestate was also determined with a set of physicochemical methods (Table 8.1).

The concentrate of organic matter, previously brought to an air-dry state by drying for 2 days in an oven at a temperature of +100°C, was also investigated. The technical conditions for digestate as a biological fertilizer were developed, as the content of fulvic (365 oD/g) and humic (600 oD/g) acids indicates its large organic component (Table 8.1). Additionally, the elemental composition of the digestate was determined by X-ray fluorescence analysis. The efficiency of the mineral component of biofertilizer is determined by the content of nutrients N/P/K – 13.7–15.0/1.3–6.6/1.5–2.1%.

As a result of chemical analysis of the substrate and digestate, their pH, HSC, BSC5, and the ratio of carbon to nitrogen content were determined. Among the lower carboxylic acids, the highest content was recorded for isobutyric acid (up to 204 mg/dm^3 in the substrate and up to 43 mg/dm^3 in the digestate). Acetic and propionic acids (84 and 92 mg/dm^3, respectively) are observed in the newly selected biomass of cyanobacteria. Among other samples, lower carboxylic acids occur in small quantities (Kvaternyuk et al. 2018b).

During a detailed study of the physicochemical and biological parameters of cyanoic methanogenesis, it was found that:

a. the presence of aliphatic organic substances in the biomass of BGA such as acetic, propionic, phenylacetic and other acids increases the yield of biogas;
b. heterocyclic compounds of non-aromatic nature (cysteine), which include nitrogen and sulfur, are a source of hydrogen sulfide formation in biogas and small amounts of ammonia;

Table 8.1 Technical requirements for biogas quality

Parameter	*Unit of measurement*	*Value*
Gas pressure	kPa	1.5–100
Minimum methane content	%	55–65
Heat of combustion	MJ/kg (kcal/kg)	≥5,500 (≥23.01)
Chlorine content	mg/nm^3 CH_4	Cl ≤ 100
Fluorine content	mg/nm^3 CH_4	F ≤ 50
Total content of chlorine and fluorine	mg/nm^3 CH_4	Cl + F ≤ 100
Silicon content	mg/nm^3 CH_4	Si ≤ 5
Sulfur content	mg/nm^3 CH_4	S ≤ 300
Hydrogen sulfide content	mg/nm^3 CH_4	$H_2S \leq 306$
Ammonia content	mg/nm^3 CH_4	$NH_3 \leq 38$
Relative humidity	%	≤60
The temperature of the gas mixture	°C	10 ≤ T ≤ 30

c. aromatic substances such as phenol, toluene derivatives, and xylene are inhibitors of biomethanogenesis;
d. biomethanogenesis occurs exponentially with optimal biogas yield under mesophilic conditions ($t = 35°C–37°C$). Increasing the temperature increases the content of impurities (CO, H_2S, NH_3) and unsaturated hydrocarbons such as ethylene, propylene, etc.

On the basis of the results of original research, a vector scheme of the technological process of processing organic matter of aquatic organisms (Figure 8.3) starts from the collection of the substrate to obtain the target products (Lugovoy et al. 2007, Kvaterniuk et al. 2018c).

The general chemical composition of substances that are the target products of the proposed biotechnology, as well as the mechanism of their production, primarily biogas mixture formed during the destruction of organic matter of aquatic organisms and the subsequent controlled process of methanogenic fermentation. The main component of biogas is methane in the amount that allows it to be used to meet the economic needs of everyday life and industrial enterprises.

The physical and chemical properties of the obtained biogas samples were determined, and a comparative analysis of the corresponding gas samples obtained from different organic substrates was performed. Approximate composition of the biogas formed by the decomposition of organic matter of aquatic organisms (based on biomass of activated sludge): methane – from 40% to 70%; carbon dioxide – from 30% to 45%; nitrogen, hydrogen sulfide, hydrogen and other gases – from 5%: up to 10%. The calorific value of biogas is from 18 to 25 MJ/m^3. Explosion hazard limits of biogas-air mixture – from 4% to 12%. The biogas mixture obtained from the organic matter of cyanobacteria is characterized by a relatively high methane content (over 73%) and virtually no hydrogen sulfide.

According to the results of the research, technical conditions (TU) were developed, which apply to "Biogas from organic matter of aquatic organisms" that establish its quantitative and qualitative characteristics. According to these specifications, biogas is used directly as fuel or for cogeneration units. 1.6–2.0 kWh can be obtained from 1 m^3 of biogas electricity or 1.6–2.9 kWh recovered thermal energy in the case of its use in a cogeneration unit. The biogas obtained from organic matter of aquatic organisms must meet the requirements listed in Table 8.1.

The technical specifications for "Digestate-based organo-mineral biofertilizer" were developed, and the quantitative and qualitative characteristics of the biofertilizer produced from the organic matter of aquatic organisms used during methanogenesis were established. In terms of the agrochemical and physicochemical parameters, the organic mixture meets the standards listed in Table 8.2.

The kinetics of biogas production was also studied. The results of the in-house studies convincingly show, as in the case of obtaining lipids from BGA, that the previous hydrodynamic cavitation was the most effective for biogas production. The obtained data are the basis for the development of technology for their processing, which involves the collection and hydrodynamic cavitation of biomass to obtain biogas and lipid extraction as raw materials for the production of biodiesel (biofuels of the third generation). The technology is protected by the relevant security documents of Ukraine (Nykyforov et al. 2016b, Chenchevoy et al. 2018).

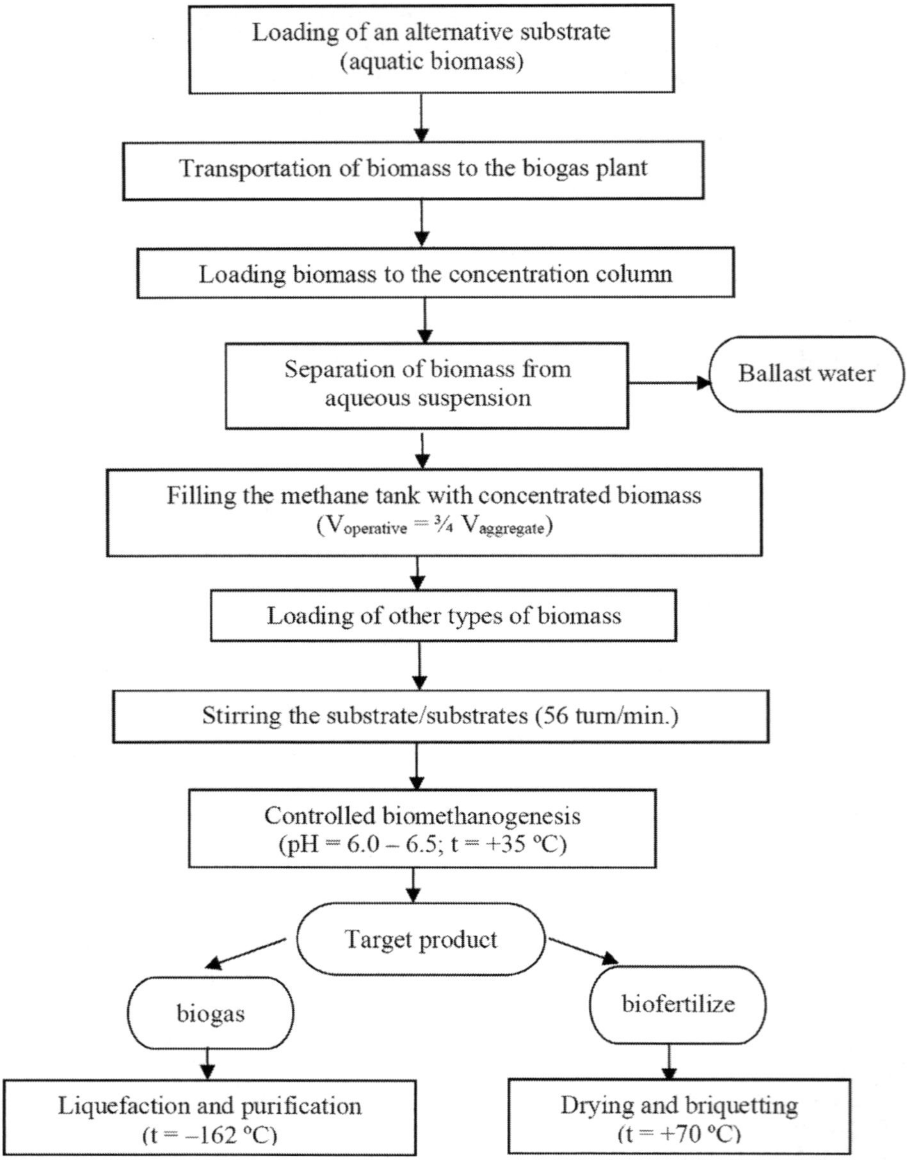

Figure 8.3 The technological process vector diagram of the organic substrate biomodification to the target product.

A mathematical model of biomethanogenesis, which takes into account various factors of methanogenesis to quantify the effects of their interaction, was presented. The standardized Pareto map allowed establishing significant factors on the amount of biogas from the BGA substrate and their mixtures with activated sludge. The intersection of standardized effects with a vertical line, which is a 95% confidence

Table 8.2 Agrochemical and physicochemical parameters of the organic mixture

Name of indicators	*Norm*	
	For use in agriculture	*For use in forestry, green building and land reclamation*
The content of fractions ≥ 50 mm, on dry matter, %, not more	2	2
Organic matter, on a dry product, %, no less	40	40
Mass fraction of moisture, %	20–95	20–95
pH, units	6.0–8.0	6.0–8.0
Nitrogen (N_xO_y) total, %	1.8	1.5
Phosphorus (P_2O_5) total, %	2.0	1.8
Potassium (K_2O) total, %	0.1	0.1

probability, indicates that the influence of factors on the response function is statically significant (Digtiar 2016b).

Graphically plotted response surfaces illustrate the dependence of the volume of biogas from the substrate BGA on variables (Figure 8.4).

The design features of the biomethanogenic plant were revealed and the calculations allowing to use the obtained practical results on an industrial scale were given. A device for increasing the density of the substrate substance – "Concentrator-digestor for the utilization of cyanobacterial biomass" was proposed and designed. The digest is a horizontally cylindrical container 2,480 mm high with a bottom diameter of 930 mm, made using the same material and methods of welded joints as the concentration column.

The installation for the collection and use of biogas includes: devices for the selection of organic matter; concentration column; methane tank (digest); storage tank for biogas (gasholder); candle for flaring excess biogas or in emergency situations (Kvaternyuk et al. 2018a).

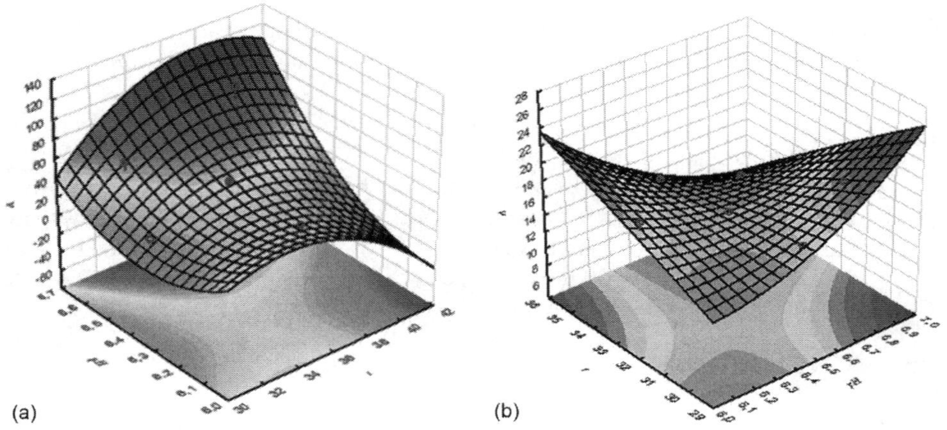

Figure 8.4 Influence of temperature and pH on the volume of biogas obtained from the substrate of cyanobacteria (a) and their mixture with activated sludge (b).

The technical specifications of the process of obtaining the biogas methane-containing mixture were developed. The potential benefits of this development for a number of industries have been identified. According to these specifications, the selection of organic matter intended for processing into biogas can be carried out from the surface of both natural and artificial reservoirs, in particular ponds and reservoirs.

Forecasting the amount of biogas produced is performed taking into account the composition and properties of raw materials, its freshness, place of selection, conditions and shelf life, pH of the aqueous extract from the substrate. The optimal mass parameters for anaerobic fermentation were also determined: humidity – from 90% to 92%; mass fraction of ash – from 15% to 16%; pH in the range of 6.0–7.5 units; the initial carbon to nitrogen ratio (C:N) should approximate 12:1.

Using these and other parameters, a virtual complex of the technological process of methane and fertilizer production from mono- and multisubstrate mixtures was created both on the basis of mass forms of aquatic organisms and with the use of traditional sources of organic matter (livestock and crop waste). The operation of the biogas plant is automated and controlled by appropriate electronic equipment and original software.

The main methods, approaches, ideas and hypotheses that determine the economic effect of the introduction of applied development in the national economy were formulated:

a. the idea of using excess biomass of aquatic organisms in general and cyanobacteria in particular as a free reducing substrate for biomethanogenesis;
b. the use of environmentally safe and cost-effective method of collecting seston (biomass);
c. the idea of using the spent substrate as an organic fertilizer;
d. the hypothesis of the possibility of inoculation of the substrate with digestate residues to reduce the duration of the first stages of biomethanogenesis;
e. the approach to improving the environment and the population by improving the quality of natural, including drinking water, due to the withdrawal of BGA from the waters of the reservoirs of the Dnieper cascade;
f. introduction of cheap biogas production and its transformation into electricity;
g. during the collection of seston in the "blooming" spots in the water area of the Kremenchug reservoir alone with an area of 2,250 km^2 in the amount of up to 50 kg/m^3 from a volume of 828 million m^3 of shallow water (depth up to 2 m; 18.4% of the reservoir area), the amount of biomass will be 4.14×10^7 tons during the growing season (120 days); by subjecting this biomass to fermentation in the process of methanogenesis, it is possible to obtain up to 108 million m3 of biogas (≈87 million m^3 of methane), which is equivalent to 76,000 tons of oil or 65,000 tons of diesel fuel, or 8,000–11,000 tons of lipids for production biodiesel.

8.3 CONCLUSIONS

On the basis of the original results of theoretical and experimental studies, the biotechnology of processing mass forms of aquatic organisms on the example of cyanobacteria to obtain a number of valuable target products was developed.

The ecological and economic significance of cyanide and prospects for efficient use of their biomass were clarified. The species composition of the initial substrate for biomethanogenesis (Microcystis aeruginosa), the biomass of which in some parts of the reservoir can reach 70–100 g/m^3, and its microbiological characteristics were determined.

The physical, chemical, and biological aspects of the process of biomethanogenesis were studied, the sequence of biochemical processes in the production of biogas from BGA was established and their features were clarified. The process of biomethanogenesis in the laboratory was modeled, as a result of which the first samples of biogas mixture from small volumes (up to 1 dm3) of substrate based on the biomass of aquatic organisms from "blooming" spots were obtained and its acceptable energy properties were proven. The chemical composition of the biogas samples of different origin (methane content up to 73%) was determined, and a comparative analysis of their physical properties was performed.

A new waste-free technological process was created, which ensures the rational use of natural resources (excess cyanide biomass) and diversifies the production of second- and third-generation biofuels. The application of the proposed biotechnology can significantly reduce the negative effects of "blooming" of natural and artificial reservoirs in the region, as well as obtain valuable energy products in the form of biomethane, biodiesel, biofertilizers, and a number of organic substances promising for use in cosmetics, pharmacology, and chemicals.

Substantiation of scientific bases of ecologically safe biotechnology of cyanide processing was proven, efficiency of using the biomass of other mass forms of aquatic organisms (higher aquatic vegetation, sea macrophytes, etc.) as substrate, and also multisubstrate mixes (leaf precipitation, activated sludge, sewage from dairy processing enterprises, and animal husbandry) was proven, in which the ratio of C:N approaches the values of 10:1–30:1.

The project of biomethanogenic installation was created and the technical conditions of processing the mass forms of aquatic organisms with reception of methane and organo-mineral fertilizer are developed. According to the designed scheme, an experimental research biogas reactor was built on the basis of the laboratory of ecological biotechnology and bioenergy of Kremenchuk National University.

The use of different substrates for obtaining biogas methane-containing mixture was studied. A critical analysis of the theoretical and practical results was made. The technical conditions for the target products of biotechnology – biogas and organo-mineral fertilizers were developed.

A mathematical model of the process of biomethanogenesis with the method of central composite rotatable planning of the complete factorial experiment CFE-2 with star points was developed. A virtual complex of the technological process of methane and fertilizer production based on mono- and multisubstrate mixtures was created, in which both mass forms of aquatic organisms and traditional sources of organic matter (agricultural and forestry waste) can be used.

The technical and economic effect of bioconversion of organic mass of cyanobacteria was calculated. The cost savings for the production of biomethane due to the substrate, as the biomass of BGA is free, is at least 42 €/t. The additional profit is due to the payment of the relevant enterprises for the processing of their waste, which is transported to biogas stations of municipal use also at the expense of the owner.

For the first time, a scientific substantiation was given for the method of obtaining methane and fertilizers from mass forms of aquatic organisms on the example of cyanobacteria, which allows applying the results to solve the problem of economically feasible use of renewable substrate, as well as during the further development of natural biotechnologies using other renewable substrates.

For the first time, the technical conditions of the process of biomethanogenesis and production of organo-mineral fertilizer on the basis of the substrate from cyanide biomass were developed, which allows solving specific socio-ecological problems caused by the "blooming" of reservoirs.

For the first time, the practical ways of utilization of digestate, including its application in agriculture and forestry, were offered.

The procedure and algorithm for calculating the capacity of an industrial plant for biomethanogenesis were improved.

The practical significance of the obtained results lies in the possibility of using the results of theoretical and practical research in biotechnology, environmental protection and alternative energy, agriculture and forestry, pharmacology and cosmetology, namely, by using cyanea biomass to obtain valuable biologically active substances and second-generation biofuels. The implementation of the results will provide an opportunity to reduce the environmental risk of the negative impact of the "blooming" of water bodies on the environment and human health, as well as reduce the anthropogenic impact on the Dnieper hydroecosystem – the main source of fresh water in Ukraine.

REFERENCES

Barsukova, T.N., Belyakova, G.A., Prokhorov, V.P. & Tarasov, K.L. 2005. *Malyi practicum po botanike. Vodorosli i griby*. Moscow: Akademiya.

Buzovsky, Y.A. Vytvycka, O.D. & Skrypnychenko, V.A. 2008. Netradytsiyni dzherela energii – vymogy chasu. *Nayk. Visn. Naz. Agrar. Un-tu* 119: 289–294.

Chenchevoy, V.V. et al. 2018. *Shlyahy optymizatsii protsesu metanogenezu z fitomasy hydrobiontiv*. XVIII Mizhnarodnoyi naukovo-praktychnoyi konferentsii "Ideyi academica V.I. Vernadskoho ta problemy staloho rozvytku osvity i nauky", Kremenchuk: KrNU.

Digtiar, S.V. 2016a. Biotehnologiya metanogenezu na osnovi biomasy tsianobacteriy. *Visnyk KrNU* 5(100): 94–99.

Digtiar, S.V. 2016b. Qualitative and quantitative characteristics of biogas of cyanea organic mass. *Environmental Problems* 1(2): 149–153.

Grebin, V.V. et al. 2014. *Vodny fond Ukrainy: Shtuchni vodoymy – vodoskhovyshcha i stavky: Dovidnyk*. Kiev: Inter-Pres Ltd.

Gregirchak, N.M. 2009. *Microbiologiya kharchovyh vyrobnytstv: Laborator*. Practicum. – K.: NUFT, 302 p.

Kustovska, O.V. 2005. *Metodologiya systemnogo pidhodu ta naukovyh doslidzhen*. Ternopil: Economichna dumka.

Kvaternyuk, S. et al. 2018a. Multispectral measurement of parameters of particles in heterogeneous biological media. *Proceedings of the SPIE* 10808: 108083K.

Kvaternyuk, S. et al. 2018b. Indirect measurements of the parameters of inhomogeneous natural media by a multispectral method using fuzzy logic. *Proceedings of the SPIE* 10808: 108082P.

Kvaterniuk, S. et al. 2018c. Mathematical modeling of light scattering in natural water environments with phytoplankton particles. *Proceedings of the 18th International Multidisciplinary Scientific GeoConference SGEM*, Vol. 18, Issue 2.1, pp. 545–552, Albena, Bulgaria.

Lemeza, N.A. 2008. *Algology and Mycology. Practical Manual.* Minsk: Vysheyshaya shcola.
Lugovoy, A.V. et al. 2007. *Sposib otrymannya biogazu z synyozelenyh vodorostey.* Patent na korysnu model No. 24106 Bulletin 9.
Nykyforov, V.V. & Kozlovs'ka, T.F. 2002. Khimiko-biologicheskiye prichiny uhudsheniya kachestva prirodnoy vody. *Visnyk KDPU* 6(17): 82–85.
Nykyforov, V.V., Kozlovs'ka, T.F. & Digtiar, S.V. 2008. Khimicheskaya biologiya methanogenesa sine-zelyonyh vodorosley i polozhytelnye effecty ih utilizatsii. *Ecologychna bezpeka* 2(2): 83–91.
Nykyforov, V.V. et al. 2011. Nature protection and energy-resource saving technology of green-blue algae utilization in Dnieper reservoirs. *Transactions of Kremenchuk Mykhailo Ostrohradskyi National University* 1(66): 115–117.
Nykyforov, V.V. & Avramenko, A.Y. 2014. Kharakterystyka suchasnogo stanu yakosti pidzemnyh i poverhnevyh vod pivdennoyi chastyny Poltavskoyi oblasti. *Visnyk KrNU* 1(84): 179–183.
Nykyforov, V.V. et al. 2016a. The biotechnological ways of blue-green algae complex processing. *Eastern-European Journal of Enterprise Technologies* 5/10(83): 11–18.
Nykyforov, V.V. et al. 2016b. *Sposib vyrobnytstva metanu ta dobryva.* Patent na korysnu model No. 104743 Bulletin 3.
Nykyforov, V.V. et al. 2018. On additional possibilities of using the blue-green algae substrate and digestate. In Sobczuk, H. et al. (ed.) *Water Supply and Wastewater Disposal*: 207–220. Lublin: Wydawnictwo Politechniki Lubelskiej.
Pastorek, Z. et al. 2006. Proizvodstvo biogaza iz smesi organicheskih materialov. *Naukovyi visnyk NUBIP Ukrayiny* 95(1): 144–149.
Perekrestov, V.I. 2014. *Praktychni metody elektronnoyi microscopiyi.* Sumy: Sumskyi Derzhavnyi Universitet.

Chapter 9

Hyaluronic Acid as a Product of the Blue-Green Algae Biomass Processing

Tetyana F. Kozlovs'ka and Marina V. Petchenko
Kharkiv National University of Internal Affairs

Olga V. Novokhatko and Olena O. Nykyforova
Kremenchuk Mykhailo Ostrohradskyi National University

Zhanna M. Khomenko
State University "Zhytomyr Politechnika"

Paweł Komada
Lublin University of Technology

Saule Rakhmetullina
East Kazakhstan State Technical University named after D. Serikbayev

Ainur Ormanbekova
Al-Farabi Kazakh National University

CONTENTS

9.1 INTRODUCTION

Nowadays, the problem of natural and technogenic-chemical pollution of natural waters surface with specific organic substances, which are accumulated in aquatic flora, fauna, and humans, is quite urgent. In this regard, considerable attention is paid to the studies aimed at preventing or limiting the mass development of blue-green algae (BGA) in artificial and natural reservoirs. The outbreaks of "blooming" increase the level of toxicity of natural waters, worsen the regime of reservoirs, as well as suppress the vital activity of aquatic organisms and residents of adjacent biotopes. The annual seasonal process of "blooming" and the subsequent death of aquatic organisms necessitate a thorough analysis of the influence of environmental conditions on the products

DOI: 10.1201/9781003177593-9

of the reservoir, mathematical modelling of eutrophication (Pasenko et al. 2016), as well as the development of technological options for solving the environmental problem (Zagirnyak et al. 2017, Nykyforov et al. 2016).

It was established that about 40 representatives of different genera of toxigenic cyanobacteria, including Microcystis, Anabaena, Nodularia, Nostoc, and Cylindrospermopsis, are sources of eutrophication. However, the main accumulator of organic matter during the "blooming" period of the Dnieper is the representative of photosynthetic cyanobacteria – *Microcystis aeruginosa* (Nykyforov et al. 2016). It accounts for up to 90% of the biomass in blooming spots – the sites of the greatest accumulation of cyanide cells in a reservoir. Beside this, the cyanobacterial actinomycete associations, as well as the *Streptomyces pluricolorescens* and *S. cyaneofuscatus*, actinomycetes are the sources of blooming natural surface waters.

Solving the problem of removal and processing of BGA will allow the targeted use of natural producers of biomass, its components, containing important food, feed, medical, pharmaceutical, perfumery, rural, and forestry components (Zagirnyak et al. 2017, Nykyforov et al. 2016). In connection with the above, this work is devoted to the possibilities of extracting hyaluronic acid (HA) from the BGA biomass and further development of biotechnology for the HA production using various actinomycete associations, including the Streptomyces actinomycetes, which are present in the biomass of BGA.

9.2 MATERIAL AND RESEARCH RESULTS

At the end of the 20th and the beginning of the 21st centuries, there was a clear tendency to introduce various bioprocesses into the industry and to replace the traditional methods of producing a number of substances of medical, cosmetic, food, and fodder purposes for the biotechnological method of production. At the same time, the disclosure of the functions and mechanisms of the biological action of a number of biopolymers contributes to the creation of new products and preparations based on them.

One of these biopolymers of animal and plant origin is HA. This glycosaminoglycan is used as part of various medical, cosmetic, and veterinary preparations. HA has found its application in surgery as a substitute for synovial fluid in joints and as a lubricant and chondroprotective component; in dermatology, it is used as a remodelling agent in the correction of the age-related deformations of the facial skin, especially the skin around the eyes, in gynaecology. HA is one of the main components of the intercellular matrix of vertebral connective tissue. It contributes significantly to cell proliferation, migration, and can also play a role in preventing tumour formation. A human body weighing 70 kg, on average, contains 15 g of HA and the third part of which is updated daily (Savoskin et al. 2017). The range of HA applications is constantly growing, as there is an increasing demand for this type of biopolymer, and, consequently, interest in alternative sources of its obtaining.

The traditional method of HA obtaining is based on the extraction of a biopolymer from various organs of mammals and birds, for example, from the vitreous body of the cattle eye, crests of chickens, or umbilical cord of newborns (Savoskin et al. 2017, Fedorishchev et al. 2000, Rjashentsev et al. 1994, Radaeva & Kostina 1998). The disadvantage is the limited raw material base for the industrial production of this polysaccharide, which cannot fully satisfy the ever-growing demand for HC. In addition, the supply of this raw material may be seasonal or uneven. Moreover, there is a risk of infection with nonspecific host-specific viruses and other infectious agents. The allocation of HA from raw animal material is often complicated by the fact that the biopolymer exists in a complex with proteins.

A biotechnological method of obtaining HA, based on the cultivation of the corresponding microorganisms-producers, is deprived of these shortcomings. The ability to synthesize HA is inherent in cultures of different cells (Streptococcus, Pasteurella, and Streptomyces) and some types of microalgae of the genus Chlorella. HA is a biotechnological product that is obtained by biological fermentation (Rjashentsev et al. 1994). The known methods for producing HA are divided into two groups: the physicochemical method, which consists in extracting HA from the tissues of animal raw materials, and the microbial method for producing HA based on producing bacteria (Rjashentsev et al. 1994, Radaeva & Kostina 1998).

It should be noted that the composition of the substrate of animal raw materials (Szwajczak 2003, Fedoryshchev 2006) and the biomass of BGA (Table 9.1) have much in common (Nykyforov et al. 2018). The allocation of HA from animal raw materials is often complicated by the fact that in the tissues and organs of mammals and birds

Table 9.1 The chemical content of the chicken combs and blue-green algae biomass

Chicken combs		*Blue-green algae biomass*	
Macronutrients		*Humid*	*dry*
Mg	46.8 mg/kg	130 mg/dm^3	105 mg/dm^3
K	393.7 mg/kg	115 mg/dm^3	120 mg/dm^3
Ca	102.3 mg/kg	70 mg/dm^3	30 mg/dm^3
Na	2500.9 mg/kg	Undefined	Undefined
P	Undefined	0.4%	6.6%
Micronutrients			
Fe	46.3 mg/kg	1.49%	1.69%
Zn	1.90 mg/kg	0.024%	0.03%
Cк	5.60 mg/kg	–	–
Mn	0.19 mg/kg	1.14%	1.46%
Co	0.1 mg/kg	–	–
Content of amino acids, vitamins and fatty acids			
Vitamins (thiamine, ascorbic and pantothenic acids)	9.1%–11.3%	95–263‰	15–72‰
Essential amino acids: valine, threonine, tyrosine	7.9%–9.1%	undefined	undefined
Total saturated fatty acids	13.76%	380 mg/dm^3	43 mg/dm^3
Humidity	87.6%	100%	12–15%
pH	6.4	–	–
Gluconic acid racemate	–	98	91
Lactose	–	21	<20
Galacturonic acid	–	25	<20
Xylose	–	<20	<20
N-acetyl-D-glucosamine	–	30	<20
Glycerin	–	36	<20

(e.g., in the crests of chickens), the biopolymer exists in a complex with proteins, proteoglycans, and, besides, the related glycosaminoglycans are often present in animal raw materials (Savoskin et al. 2017).

The microbiological method for HA production based on the producer of Streptomyces makes it possible to relatively easily purify it and fractionate by molecular weight in the absence of HA-related proteins. Therefore, biotechnological HA preparations have higher quality indicators. As it can be seen from the data in Table 9.1, the chemical composition of the biomass of BGA is more suitable for the production of HA in comparing to the animal sources. The biomass of BGA contains carbon as the main skeleton for building the skeleton of HA, glucose, and has a high content of proteins, including enzymes (23.0%–82.6%), carbohydrates (6.6%–70.0%), and lipids (2.0%–12.0%), which include polyunsaturated fatty acids and vitamins (B_1 (thiamine), B_2 (riboflavin), B_3 (vitamin PP – nicotinamide), B_6 (pyridoxal, pyridoxine), B_7 (biotin), B_9 (folic acid), C, A, E, and K) and minerals (iron, magnesium, calcium, iodine Boron, zinc, and copper) (Fedoryshchev 2006, Nykyforov et al. 2018, Syrenko & Kozitskaya 1988). All of these components are catalysts for the allocation of HA.

Our attention was focused on comparing the technology of obtaining HA from animal raw materials and the possibility of using leachate (liquid digestate) after separation of the solid phase of digested biomass of cyanobacteria.

Under laboratory conditions, studies were carried out of digestate of biomass BGA for the presence of bacteria of the genus Actinomyces, acting as a source of HA. In water that remains after the separation of the raw biomass of the BGA, these bacteria were detected. After bio-methanogenesis, they are absent, since under the conditions of methanogenic fermentation, the bulk of Actinomyces, including Streptomyces, die. They are aerobic heterotrophs, which need nutrients and molecular oxygen. The anaerobic conditions of methanogenesis transform all components of the synthesis of HA into methane, carbon dioxide, and hydrogen (Zagirnyak et al. 2017).

In addition, the possibilities of obtaining HA were studied after the extraction of lipids from the biomass of BGA (Malovanyy et al. 2019a), as well as after its cavitation treatment (Malovanyy et al. 2019b). As expected, it was not possible to extract HA. There are several reasons for this. First, HA forms complexes with lipids, and after their extraction by the cavitation method or laser treatment, not only the complexes are destroyed but also the chemical bonds of the biopolymers with the formation of glucose aminooglycan mono- and dimolecules. The direct cavitation treatment of the BGA biomass destroys not only organic biopolymers but also molecules of saturated organic fatty acids, which take part in the stabilization of the complexes "hyaluronic acid-lipid". In the process of electrochemical treatment of the BGA biomass (Nykyforov et al. 2019), subsequent studies did not reveal the presence of HA either.

It should be noted that cyanobacterial actinomycete associations and actinomycetes, which are present in small amounts in natural waters, play an important role in the synthesis of HA. They enter the water from the places of primary soil formation along the underground aquifers from sedimentary carbonate rocks, where actinomycetes are involved in the stabilization of the bacterial block of the system and enhance the photosynthetic activity of algae (Matseliukh 2003, Zenova & Zviahyntsev 2002, Ignatova & Gurov 1990). Obviously, actinomycetes, as associative symbionts, have a positive effect on the ecosystem as a whole, due to the stimulating effect on the nitrogen-fixing ability of cyanobacteria and increased protection of the entire system from pathogenic microorganisms due to the release of antibiotics as waste products.

HA – $(C_{14}H_{21}NO_{11})_n$ – is an organic compound belonging to the group of non-sulfated glucose aminoglycans (Figure 9.1). The presence of numerous sulfated groups in related glucose aminoglycans is the cause of numerous isomerism, which is not observed in gialuronic acid, which is always chemically identical, regardless of the methods and sources of production.

The HA molecule is characterized by the formation of internal intermolecular hydrogen bonds both inside the molecule and between adjacent carbon residues, which are at a considerable distance from each other, and in an aqueous solution even between neighbouring molecules through a carboxyl and acetamide group. The HA solutions have a pH < 7 due to the carboxyl group. The acidic properties of hyaluronate make it possible to obtain water-soluble salts with alkali metals. HA is an anionic linear polysaccharide with different molecular weights (10^5–10^7 Da). The molecular weight depends on the method of preparation; the resulting HA is always chemically identical to natural, due to the absence of isomerism (Zenova & Zviahyntsev 2002).

HA is an amorphous polymer. The purified preparation is a white fine powder. HA macromolecules have a linear structure and are characterized by a high degree of asymmetry. The molecular weight of HA depends on the method of preparation and may be 50–8,000 kDa. HA is a hydrophilic polymer and is characterized by high sorption ability to water molecules. In the presence of water, HA forms elastic, elastic (soft) gels, while the binding 10,000-fold volume of water. HA, due to its hydrophilic properties, regulates the water balance of the tissue. High hydrophilicity determines the widespread use of polysaccharide in cosmetology.

HA and its derivatives are massive polysaccharide biopolymers consisting of repeating simple units – disaccharides. In a complex molecule of HA, alternate acetylglucosamine and glucuronic acid are structural components of the polymer. HA is synthesized by a class of embedded membrane proteins, that is, enzymes called hyaluronate synthetases.

Hydrolytic destruction of HA. With complete acid hydrolytic cleavage of HA, β-D-glucuronic acid, β-N-glucosamine, and acetic acid are formed (Figure 9.2) (Tan et al. 1990).

The enzymes from the class of hyaluronidases cause biodegradation of HA. They are divided into two main types: testicular and bacterial hyaluronidases. The products of enzymatic degradation of HA under the influence of heat-resistant testicular hyaluronidase are tetraoligosaccharides which consist of two elementary units of a macromolecule of HA, connected by a β-(1–4)-glycoside bond, while under the action

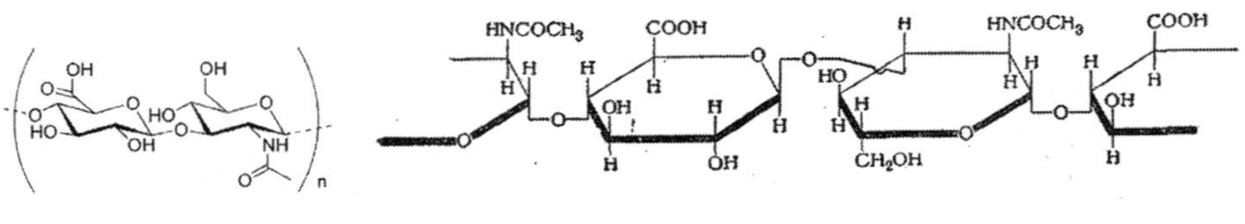

Figure 9.1 The chemical formula of hyaluronic acid (Savoskin et al. 2017).

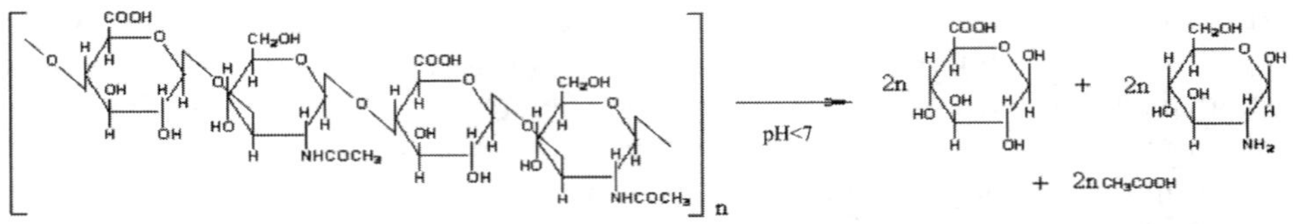

Figure 9.2 Hydrolytic destruction of hyaluronic acid (Savoskin et al. 2017).

of bacterial hyaluronidase – the disaccharides: (β-N-glucosamine) β-D-glucuronates are formed (Figure 9.3) (Savoskin et al. 2017).

HA is an amorphous polymer. The purified preparation is a white fine powder. HA macromolecules have a linear structure and are characterized by a high degree of asymmetry. The molecular weight of HA depends on the method of preparation and may be 50–8,000 kDa. Presumably, natural HA in its native state has a molecular weight of 1,000–20,000 kDa. For example, the average molecular weight of the polysaccharide contained in human synovial fluid is 3,000–3,500 kDa (Syhaeva et al. 2012, Sutherland 1998).

HA is a hydrophilic polymer and is characterized by high sorption ability for water molecules. In the presence of water, HA forms elastic (soft) gels as an effect of binding water in a volume of 10,000-fold higher than HA.

HA dissolves in water and in an aqueous solution of sodium chloride, does not dissolve in organic solvents, which was taken as the basis for the extraction of HA from the digestate of CZV biomass. Experimental studies were carried out in two directions:

- extraction of HA directly from the biomass of BGA;
- production of HA from digestate BGA.

In both cases, the experiments were performed under the same conditions: 500 cm^3 samples were taken. In the first case, the samples were sedimented for 120 hours to completely precipitate biomass. The samples were filtered and acetate buffer was added to maintain the medium acidity and 2 dm^3 of sodium chloride solution of the appropriate concentration (Table 9.2). In the second case, acetate buffer and 2 dm^3 of sodium chloride solution of the corresponding concentration were also added directly to the digestate.

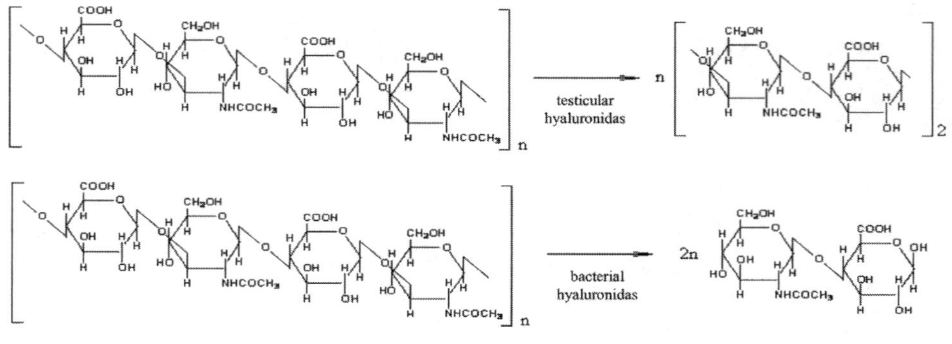

Figure 9.3 Enzymatic biodegradation of hyaluronic acid (Savoskin et al. 2017).

Table 9.2 The results of obtaining hyaluronic acid from digestate biomass BGA

C NaCl, %	*Hyaluronic acid mass, 10^{-3} g*									
	Formation time of hyaluronic acid, hours									
	12	*16*	*20*	*24*	*28*	*30*	*36*	*48*	*60*	*72*
2	3.72	3.77	3.82	3.9	3.92	3.96	4.01	4.05	4.08	4.12
5	4.16	4.19	4.23	4.26	4.29	4.33	4.36	4.39	4.42	4.45
8	4.48	4.5	4.53	4.56	4.59	4.61	4.64	4.67	4.69	4.72
12	4.75	4.77	4.8	4.82	4.85	4.87	4.9	4.92	4.95	4.97
15	5	5.03	5.05	5.08	5.1	5.13	5.15	5.18	5.2	5.23
18	5.25	5.28	5.31	5.33	5.36	5.39	5.41	5.44	5.47	5.5
20	5.52	5.55	5.58	5.61	5.64	5.67	5.71	5.74	5.77	5.81
25	3.16	4.2	4.8	3.9	5.17	4.9	4.65	4.7	4.8	5.1

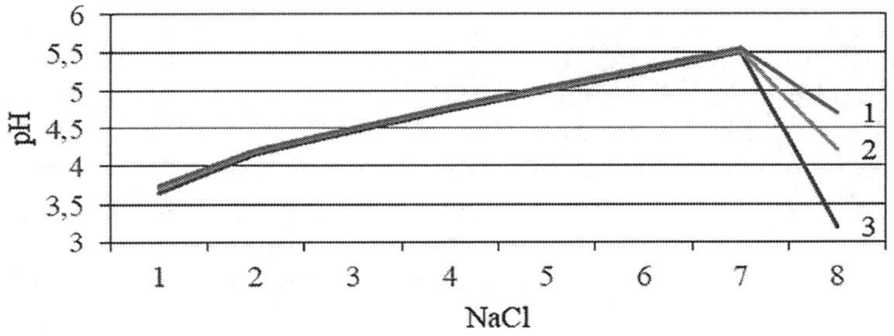

Figure 9.4 The dependence of hyaluronic acid formation on the concentration of sodium chloride over time: 1–12, 2–16, and 3–24 hours.

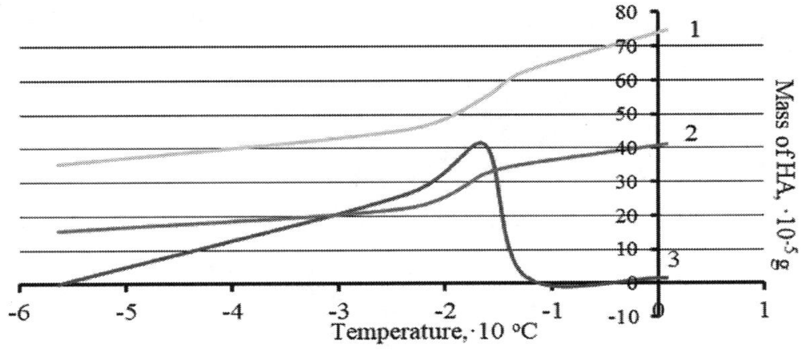

Figure 9.5 The dependence of hyaluronic acid formation on the deposition temperature of sodium chloride solution: 1%–20%, 2%–15%, and 3%–25%.

After 5–6 hours, the onset of turbidity of the solution was observed, which intensified when the samples were kept at low temperatures (from 0°C to −60°C). Gradually, sedimentation was observed at the bottom of the gel-like substance, slightly yellowish in colour (Figures 9.4 and 9.5).

Identification of the substance was carried out by comparison with the preparation of HA used in cosmetology. Thin-layer chromatography in the methanol: water: butanol system (1:4:1) showed close *R* values in the range of 0.65–0.71. The obtained HA is not a pure product, since it can be a lipid conglomerate with capsules of HA, coordination complexes with water molecules due to its high sorption capacity, with organic substances of different structures that are present in natural waters as algae vital products. Since HA is capable of forming salts with metal ions, an attempt was made to isolate HA in the form of the corresponding salts with solutions of potassium chloride and sulphate, as well as calcium chloride (Figure 9.6). It was found that HA cannot be isolated with potassium ions. Solutions of calcium chloride with a concentration of 5%–20% are more optimal. In this case, the consistency of the resulting HA changes to partially powdery, although gelability also remains.

As can be seen from the results obtained, the cooling temperature of the solution significantly affects the formation of HA. Temperatures above zero, especially above 10°C, contribute to the breaking of oxygen-hydrogen bonds in the biopolymer and intermolecular hydrogen bonds. As a result, di- and monoglucose-aminoglycans are formed (Figure 9.3).

A sharp drop in the yield of HA in the temperature region of −18°C at a concentration of sodium chloride solution of 25% cannot be clearly explained (Figures 9.4 and 9.5). It is possible that the so-called "glass transition" of molecules, characteristic of natural biopolymers, plays a role. At lower concentrations of sodium chloride solution, this effect is not observed; therefore, the yield of HA also increases.

When sodium chloride is replaced by calcium chloride, the picture changes dramatically. With an increase in the solution concentration from 15% to 25%, the yield of HA increases as well (Table 9.2). The nature of the curves can be described by a dependence of the type

$$m = \pm aC^2 \pm bC + c, \tag{9.1}$$

with the accuracy of the approximation, $R^2 = 0.97–0.98$. In the expression (9.1), C is the concentration of the calcium chloride solution.

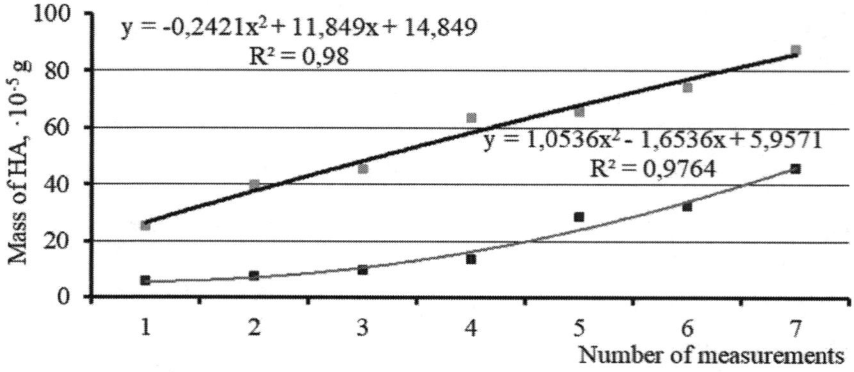

Figure 9.6 The dynamics of hyaluronic acid yield in calcium chloride solution.

The stability of the calcium salts of HA can be explained in terms of the energy of the Ca–O ionic bond and the change in the spatial configuration of glucose rings. However, this point requires further research, since a change in the spatial arrangement of structural fragments can affect biological activity – strengthen or level it.

9.3 CONCLUSIONS

The possibility of extracting HA from the BGA biomass which is a component of natural surface waters during periods of intensive blooming, as also the digestate obtained after methane fermentation of this biomass was experimentally confirmed. It was shown that cyanobacteria of natural waters: *Microcystis aeruginosa* and cyanobacteria and actinomycete associations of *A. variabibilis*, *S pluricolorescens* and *S. cyaneofuscatus* are the main sources of HA. Along with this, at the stage of extraction of HA, no technological equipment is required.

The possibility of extracting HA was experimentally confirmed. The dynamics of its quantitative characteristics was presented. The studies to establish the dependence of the extraction of HA on the concentration of solutions of sodium and calcium chlorides were carried out. It was shown that the optimum conditions for the use of sodium chloride in the precipitation of HA is a concentration of 15%–20% at a temperature of the reaction medium of −5°C to −18°C. The use of calcium chloride solutions increases the yield of HA by 1.5–1.7 times in the temperature range of 0 to −10°C.

The obtained samples of HA were compared with its true solutions used in cosmetology using thin-layer chromatography. In the process of processing biomass of BGA, it was found that the processes of methanogenesis destroy molecules of not only HA but also the simplest glucose aminoglycans. In order to increase the yield of HA from the biomass of BGA, neither its laser processing with the aim of extracting lipids nor cavitation is suitable. This is due to the weakness of chemical bonds in the glucose aminoglycanate bridges of the HA molecule itself.

REFERENCES

Fedorishchev, I.A., Chernyshov, A.A. & Epifanov, A.E. 2000. *Method of preparing hyaluronic acid.* Russia patent: 2157381; Priority date: 01.03.1999; Publication info: 10.10.2000.

Fedoryshchev, Y.A. 2006. *Biophysical characteristics of modified hyaluronic acid.* Tula: Ph.D. dissertation.

Ignatova, E.Y. & Gurov, A.N. 1990. Extraction principles and purification of hyaluronic acid. *Khimiko-Farmatsevticheskii Zhurnal* 3: 42–46.

Malovanyy, M., Moroz, O., Hnatush, S., et al. 2019a. Perspective technologies of the treatment of the wastewaters with high content of organic pollutants and ammoniacal nitrogen. *Journal of Ecological Engineering* 20(2): 8–15.

Malovanyy, M., Zhuk, V., Nykyforov, V., Bordun, I., Balandiukh, I. & Leskiv, G. 2019b. Experimental investigation of *Microcystis aeruginosa* cyanobacteria thickening to obtain a biomass for the energy production. *Journal of Water and Land Development* 43(1): 113–119.

Matseliukh, A.B. 2003. Streptomycetes – producers of polyketide antibiotics. *Microbiological Journal* 1–2(65): 168–181.

Nykyforov, V., Malovanyy, M., Aftanaziv, I. et al. 2019. Developing a technology for treating blue-green algae biomass using vibro-resonance cavitators. *Naukovyi Visnyk Natsionalnoho Hirnychoho Universytetu* 6: 181–188.

Nykyforov, V., Malovanyy, M., Kozlovs'ka, T., et al. 2016. The biotechnological ways of blue-green algae complex processing. *Eastern-European Journal of Enterprise Technologies* 5/10(83): 11–18.

Nykyforov, V.V., Kozlovs'ka, T.F., Novokhatko, O.V. & Digtiar, S.V. 2018. On additional possibilities of using the blue-green algae substrate and digestate. In H. Sobczuk & B. Kowalska (eds.) *Water Supply and Wastewater Disposal: Monograph*: 207–220. Lublin: Politechnika Lubelska.

Pasenko, A.V., Novokhatko, O.V., Kozlovs'ka, T., Digtiar, S. & Nikiforova, O. 2016. The main approaches of mathematical modeling of biological productivity of cyanobacteria as a raw-material base for bioconversion. *Ecological Safety* 2(22): 118–127.

Radaeva, I.F. & Kostina, G.A. 1998. *Method for production of hualyronic acid*. Russia patent: 2102400; Priority date: 08.08.1995; Publication info: 20.01.1998.

Rjashentsev, V.J., Nikol'skij, S.F., Vajnerman, E.S., Poljakov, V.I., Gurov, A.N., Ovchinnikov, A.N. & Ignatova, E.J. 1994. *Method of hyaluronic acid preparing*. Russia patent 2017751; Priority date: 22.05.1991; Publication info: 15.08.1994.

Savoskin, O.V., Semyonova, E.F., Rashevskaya, E.Y., et al. 2017. A description of different methods used to obtain hyaluronic acid. *Scientific Review. Biological Science* 2: 125–135.

Sutherland, I.W. 1998. Novel and established applications of microbial polysaccharides. *Trends in Biotechnology* 16(1): 41–46.

Syhaeva, N.N., Kolesov, S.V., Nazarov, P.V. & Vyldanova, R.R. 2012. Chemical modification of hyaluronic acid and use in medicine. *Bulletin of the Bashkir University* 17(3): 1220–1241.

Syrenko, L.A. & Kozitskaya, V.N. 1988. *Biologically Active Algae Substances and Quality of Water*. Kiev: Nauka Dumka.

Szwajczak, E. 2003. Dependence of hyaluronan aqueous solution viscosity on its microstructure. Part I. *Russian journal of biomechanics* 7(3): 87–98.

Tan, S.W., Johns, M.R. & Greenfield, P.F. 1990. Hyaluronic acid – a versatile biopolymer. *Australian Journal of Biotechnology* 4(1): 38–43.

Zagirnyak, M.V., Nykyforov, V.V., Malovanyi, M.S., et al. 2017. *Ekolohichna biotekhnolohia pererobky synio-zelenykh vodorostei*. Kremenchuk: PP Shcherbatykh O.V.

Zenova, H.M. & Zviahyntsev, D.H. 2002. *Diversity of Actinomycetes in Terrestrial Ecosystems: A Textbook*. Moscow: MFU.

Chapter 10

Prospects for the Use of Cyanobacterial Waste as an Organo-Mineral Fertilizer

Myroslav S. Malovanyy, Ivan S. Tymchuk, and Christina M. Soloviy
Lviv Polytechnic National University

Olena O. Nykyforova
Kremenchuk Mykhailo Ostrohradskyi National University

Dmytro V. Cherepakha
Vinnytsia National Technical University

Waldemar Wójcik
Lublin University of Technology

Indira Shedreyeva and Gayni Karnakova
M. Kh. Dulaty Taraz Regional University

CONTENTS

10.1 INTRODUCTION

Insufficiently treated municipal effluents (Sakalova et al. 2019), organic pollutants (Zelenko et al. 2019) and landfill leachates (Malovanyy et al. 2018, 2019) enter surface water bodies and cause an increase in the concentrations of nitrogen and phosphorus compounds, which causes uncontrolled development of cyanobacteria and eutrophication of surface water bodies. The problem of water eutrophication is extremely important today. Every year, a large area of reservoirs blooms killing many organisms that live in these reservoirs. Water blooms occur due to the mass development of cyanobacteria, that is, blue-green algae. Most often, this process occurs in fresh standing water, but is also possible in sea water. The water becomes

DOI: 10.1201/9781003177593-10

green, yellow-brown, and sometimes red, depending on the algae. The harm of water blooming for wildlife is the production of strong toxins that are dangerous to both humans and animals, water quality reduction, loss of useful qualities and properties of the aquatic ecosystem, as well as blooming water disrupts the esthetic appearance of the reservoir and spreads odor.

The studies conducted on the example of the Kremenchug reservoir showed that the forced removal of excess organic matter from aquatic ecosystems with its subsequent bioconversion can partially solve or at least reduce the severity of the environmental problems associated with the "blooming" of reservoirs. The wide range of applications of excess biomass of cyanobacteria formed during the "blooming" of water bodies is mainly due to the presence in its chemical composition of the components that can be used directly in many industries and involved in biotechnological processes. In particular, the most promising is the involvement of concentrated organic matter to obtain a biogas mixture.

Relevant technologies have already been tested in the laboratory, as evidenced by the relevant patents and a number of publications in domestic and foreign scientific journals (Cavinato et al. 2017, Malovanyy et al. 2016). However, this process is not always cost-effective and at the end there is still a significant amount of spent biomass. Therefore, the reduction of waste biomass and finding ways to use it in economic activities is a very important task.

After analyzing the literature, the authors came to the conclusion that the most promising use of cyanobacterial waste after biogas production is their use in agriculture as biofertilizers, and their subsequent production.

In general, cyanobacteria are well known for their ability to fix the atmospheric nitrogen (N_2) either through free residence or through symbiotic associations (Singh et al. 2019). A number of studies also showed that the use of cyanobacteria from different cultures has a positive effect on their germination, growth, and yield (Saadatnia and Riahi 2009, Osman et al. 2010). In addition, cyanobacteria provide additional application to crops up to 20–30 kg N/ha. Studies (De Caire et al. 2000, Pandey et al. 2005, Obana et al. 2007) show that the use of cyanobacteria improves the soil structure and can be used in the reclamation of acidic and alkaline soils by lowering the pH. Thus, the use of cyanobacterial cells gives us a simple, cheap, and effective slow-release biofertilizer to increase crop productivity and rehabilitate degraded land in the regions where very little or no chemical fertilizers are usually used (Singh et al. 2019).

10.2 MATERIALS AND METHODS

The primary task of the study was to check the processed biomass of cyanobacteria for the presence of heavy metals and dangerous toxins, as this is the main limiting factor for fertilizers. In order to do this, an EXPERT 3L X-ray fluorescence analyzer was used, the general view of which is presented in Figure 10.1.

The purpose of the EXPERT 3L X-ray fluorescence analyzer is measurement of mass fraction (%) of the main chemical elements by the method of X-ray fluorescence analysis. The range of measured chemical elements (control range) is from magnesium (12 Mg) to uranium (92 U). In the process of interaction of the sample with

Figure 10.1 X-ray fluorescence analyzer EXPERT 3L.

high-energy X-rays, part of the radiation passes through the sample, part is scattered, and part is absorbed by the substance of the sample. The absorption of X-rays by a substance leads to the appearance of several effects at once, one of which is X-ray fluorescence – the emission of a substance by secondary X-rays. The EXPERT 3L analyzer implements the technique of energy-dispersion X-ray fluorescence elemental analysis through the method of fundamental parameters with damage of the characteristic radiation of sample atoms by photons of the brake spectrum of a low-power X-ray tube and registration of this radiation by a semiconductor PIN detector with a thermal detector.

Further testing of the processed biomass for the presence of basic nutrients, as well as macro-and micronutrients in the forms available to plants was carried out in the state certified Laboratory of Agrochemical, Toxicological and Radiological Researches on Soil Ecological Safety and Product Quality Lviv branch State Institution "Soils Protection Institute of Ukraine".

10.3 RESULTS AND DISCUSSION

In order to establish the possibility of using processed biomass after synthesis of biogas from it as organic fertilizers, the elemental composition of dried spent biomass was determined on an EXPERT 3L X-ray fluorescence analyzer (Table 10.1). This indicator is extremely important because ultimately it determines the effectiveness of the fertilizer, the degree of balance of macronutrients in plant nutrients, and the micronutrient composition. According to the Ukrainian legal documents (TU 24.1-14005076-065-2003 "Foreign phosphorites"), the main limiting compounds in raw materials for fertilizer production are lead (Pb), cadmium (Cd), and arsenic (As).

The results of our research indicate the safety of these types of waste relative to the elemental composition, and the presence of a significant number of nutrients. Spent substrate contains more calcium and sulfur than other elements (these elements are trace elements necessary for a balanced plant nutrition), the application of which

Table 10.1 Elemental composition of dried processed biomass of cyanobacteria

Element	*Quantity, %*	*Element*	*Quantity, %*	*Element*	*Quantity, %*
$_{14}Si$	4.432 ± 0.086	$_{22}Ti$	0.081 ± 0.019	$_{34}Se$	0.007 ± 0.002
$_{15}P$	7.160 ± 0.131	$_{25}Mn$	1.139 ± 0.017	$_{35}Br$	0.053 ± 0.002
$_{16}S$	11.713 ± 0.101	$_{26}Fe$	1.492 ± 0.015	$_{38}Sr$	0.029 ± 0.002
$_{17}Cl$	8.461 ± 0.079	$_{28}Ni$	0.023 ± 0.002	$_{40}Zr$	0.004 ± 0.002
$_{19}K$	20.197 ± 0.060	$_{29}Cu$	0.006 ± 0.001	$_{46}Pd$	0.008 ± 0.002
$_{20}Ca$	45.131 ± 0.112	$_{30}Zn$	0.024 ± 0.001	$_{51}Sb$	0.025 ± 0.004

in the composition of fertilizers is appropriate. The contents of phosphorus and potassium – the main elements of plant nutrients – are at the level of the best sorts of mineral fertilizers. However, with this method, we cannot determine the form in which the compounds are and whether it is available to plants, nor can we determine the amount of one of the most important elements for plant growth and development, for example, nitrogen.

The negative factor is considerable content of chlorine, but it is often included in the form of chlorides in potassium fertilizers, which are widely used in agriculture, so its content in organic fertilizers from waste biomass is acceptable. In addition to a small amount of ballast silicon, the new potential fertilizer additionally contains microelements, that is, iron and manganese, which are necessary to ensure balanced plant development. This composition is acceptable for the use of spent biomass of cyanobacteria as fertilizer.

The analysis of the cyanobacteria biomass before and after anaerobic fermentation (Table 10.2) showed that this biomass contains a significant part of nitrogen compounds, and their number after fermentation only increases. The content of organic compounds naturally decreases after fermentation, but is quite high as for organic-mineral fertilizers. Neutral pH does not limit the practical application of these types of fertilizers on any type of soil. One of the very important factors is the presence of fulvic and humic acids, which are components for the reproduction of soil fertility, as well as plant growth stimulants in agriculture.

Table 10.2 The results of studies of fresh and processed biomass of cyanobacteria

Sample	*Dry cyanobacteria biomass*	*Dry processed cyanobacteria biomass*	*Fresh cyanobacteria biomass*	*Processed cyanobacteria biomass*
Water content, %			99.0	99.2
COD, g/kg			10.43	9.43
BOD_5, mg O_2/l			1750	530
pH (water)	7.1	7.5		
$N_{Kjeldahl}$	9.1 mg/kg	10.0 mg/kg	900 mg/l	800 mg/l
NH_4–N ($CaCl_2$)	3.4 mg/kg	6.8 mg/kg	340 mg/l	550 mg/l
NO_3–N ($CaCl_2$)	0	0	1 mg/l	2 mg/l
C_{org}, %	40.1	37.1		
C/N	4	4		
Fulvic acids	95 oD/g	65 oD/g	745 oD/l	365 oD/l
Humic acids	90 oD/g	110 oD/g	680 oD/l	600 oD/l

In further research, we determined how many nutrients, macro- and micronutrients, are in the forms available for plant growth and development (Table 10.3).

In terms of dry matter in this waste biomass, the greatest value may be:

1. Ash, 12.6%;
2. Nitrogen total, 6.36% (ammonium, 5.9%);
3. Phosphorus total, 2.1%;
4. The presence of microelements, especially mobile sulfur, which promotes the efficient absorption of nitrogen in plants.

The spent biomass of cyanobacteria, if pre-dehydrated, can be used as an organic-mineral fertilizer, because it contains a combination of organic and mineral components. The origin and composition of waste biomass can be promisingly used in organic farming to obtain environmental-friendly products, as it does not contain hazardous and toxic substances, or foreign (chemical) ballast substances. A balanced combination of a significant amount of nitrogen with phosphorus and sulfur will allow its effective assimilation by plants because this complex is necessary for the growth and development of almost all plants. A wide range of the presence of even a small amount of microelements will contribute to better provision for plant growth and development.

Table 10.3 Determination of the main agrochemical indicators of biomass cyanobacteria processed

Names of indicators	*Units*	*Actual value*		*Normative document for test methods*
		Dry matter	*Natural humidity*	
Acidity				
pH salt	pH		5.9	
pH water	pH		6.1	
Humidity	%	–	96.1	
Ash	%	12.6		
Phosphorus total	%	2.1	0.08	
Potassium total	%	0.8	0.03	
Nitrogen total	%	6.36	0.25	
Nitrogen ammonium	%	5.90	0.23	GOST 26712-85
Calcium (as plant)	%	1.68	–	GOST 26718-85
Calcium (as soil)	mg/kg	600		
Magnesium (as soil)	mg/kg	360		
Sulfur mobile (in soil)	mg/kg	21.2		GOST-26490-85
Microelements				
Copper (Cu)	mg/kg	3.2		
Zinc (Zn)	mg/kg	5.86		
Manganese (Mn)	mg/kg	77.8		
Cobalt (Co)	mg/kg	1.95		MU by atomic absorption determination
Boron (B)	mg/kg	2.22		GOST 10.154-83

Therefore, the biogas technology allows obtaining natural biofertilizer, which contains biologically active substances and microelements, in the shortest possible time by anaerobic fermentation. The main advantage of biofertilizers over traditional ones is the shape, availability, and balance of all nutrients, high level of humification of organic matter, which serves as a powerful energy material for soil microorganisms, so after application in the soil, nitrogen-fixing, and other microbiological processes are activated. This creates a positive effect on soil fertility and improves the physical and mechanical properties of the soil. The use of biofertilizers for growing crops will reduce the use of chemical fertilizers that have a negative impact on the quality and fertility of soils.

Experimental data indicate the practical feasibility and economic feasibility of using the organic mass of cyanobacteria extracted from blooming spots in the waters of the Dnieper reservoirs for industrial biogas production with subsequent use of spent substrate as a balanced organic-mineral fertilizer in forestry and agriculture.

In our opinion, further research should be conducted on the optimum conditions for the use of spent biomass of cyanobacteria for various crops in specialized research institutions of agricultural profile. In addition, research is needed to determine the optimum appearance, in order to minimize the ability to stick and caking, absence of pungent odors, convenient for transportation and application to the soil form, since the use in this state will limit both the consumer and the manufacturer. In our opinion, this is possible in the case of combining spent biomass with natural dispersed sorbents (natural zeolites, glauconites, or paligorskites). It is also possible to use this type of waste for the biological reclamation of anthropologically disturbed lands, namely to create a nutrient substrate for the biological stage of reclamation. Referring to the results of previous studies (Shkvirko et al. 2018, 2019), we have developed a theoretical scheme for the use of this type of waste in the process of creating a substrate (Figure 10.2), and usually this hypothesis requires detailed laboratory and later field research. In this way, we will be able to significantly expand the scope of use of this waste (Kvaternyuk et al. 2017).

We also conducted the research aimed at determining the most favorable concentration of processed substrate of blue-green algae for use in agriculture. In order to achieve this goal, the toxicity levels of different substrate concentrations were determined by biotesting (Petruk et al. 2015).

Two types of cultivated plants were used to determine the sowing qualities (degree of germination) of seeds as one of the criteria for assessing the possibility of using cyanobacteria biomass after methanogenesis: durum wheat – *Triticum durum L.* (monocotyledons) and sowing peas – *Pisum sativum L.* (dicotyledons) (Petruk et al. 2012).

The spent biomass of cyanobacteria diluted with distilled water was used for the studies, the experiments were performed twice (changing the temperature regime) to obtain reliable data, and samples were repeated in each of the experiments.

Germination was performed in Petri dishes using waste in different dilutions (1:10, 1:50, 1:100, 1:200). Germination was determined in the percentage of plants germinated from 100 seeds compared with the control (bidistillate) in three replicates. The results of the research are shown in Figure 10.3. During this period, the temperature was +25°C and pH = 6.0.

Detailed analysis of studies results pertaining to the test cultures' germination under the influence of substrates with different concentrations revealed the reasons that affect the similarity of these test objects:

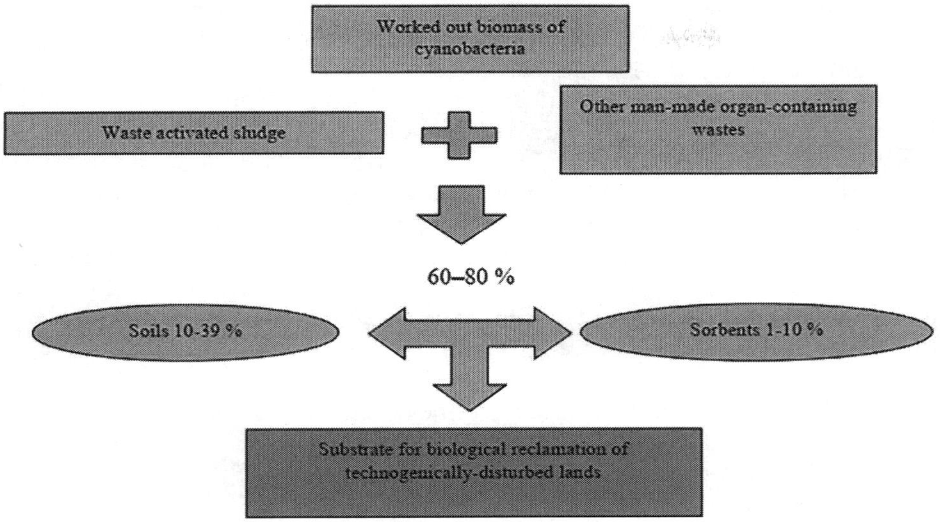

Figure 10.2 The scheme of use of the processed biomass at creation of a substrate for biological reclamation.

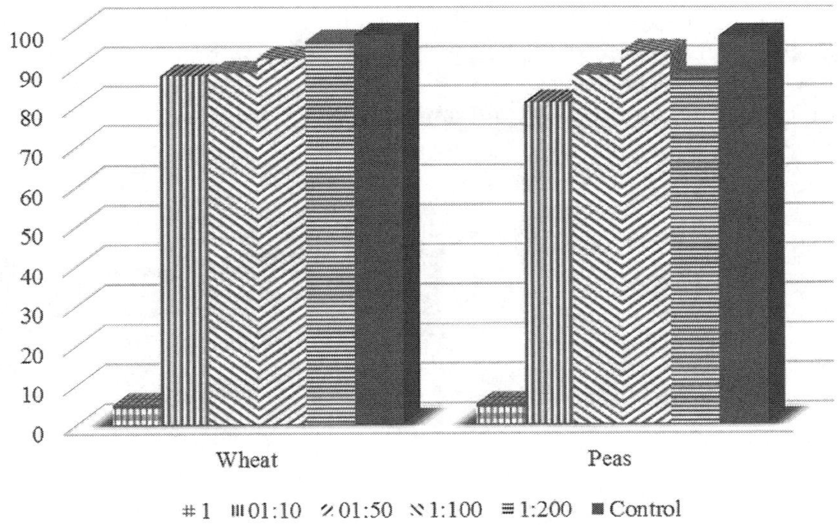

Figure 10.3 Similarity (%) of cultivated plants with the addition of processed substrate in different dilutions in the first experiment.

- germination of the studied cultivated plants in double-distilled water (control) is – 90% and 88% for wheat and peas, respectively;
- wheat germination at substrate dilutions of 1:10 and 1:50 decreases by 1.7% and 1.0%, and at 1:100 and 1:200 – increases by 5.6% and 7.0%, respectively;

- germination of peas at dilutions of the substrate 1:10 decreases by 7%, 1:50 – does not change, 1:100 and 1:200 – increases by 6.0% and 1.0%, respectively;
- optimum for use as a bioorganic fertilizer is the dilution of the processed substrate 1:200 for wheat and 1:100 for peas.

During the second germination, the conditions were changed: the temperature was 26.5°C and pH = 8.0. Germination was performed in Petri dishes using the substrate in different dilutions (1:10, 1:50, 1:100, 1:200), as well as additionally – 1:500 and 1:1000. Germination was determined in the percentage of plants germinated from 100 seeds compared with the control (bidistillate) in two replicates. The research results are shown in Figure 10.4.

Detailed analysis of studies results pertaining to the test cultures' germination under the influence of substrates with different concentrations revealed the reasons that affect the germination of seeds:

- germination of wheat at dilutions of the substrate 1:10 and 1:200 is higher compared to the control;
- germination of peas at a dilution of 1:50 decreases by 1.5%, and at 1:200 – by 3% compared to the control;
- optimum for use as a bioorganic fertilizer is the dilution of the processed substrate 1:200 – for wheat, and 1:100 – for peas.

Similarity of test objects at lower substrate concentrations (1:500, 1:1000):

- for wheat increased by 10% and 3%, respectively, compared to bidistillate;
- for peas decreased by 9% and 19.5%, respectively, compared to the control.

Thus, the optimum for use as a bioorganic fertilizer is a dilution of the processed substrate 1:500 for both crops.

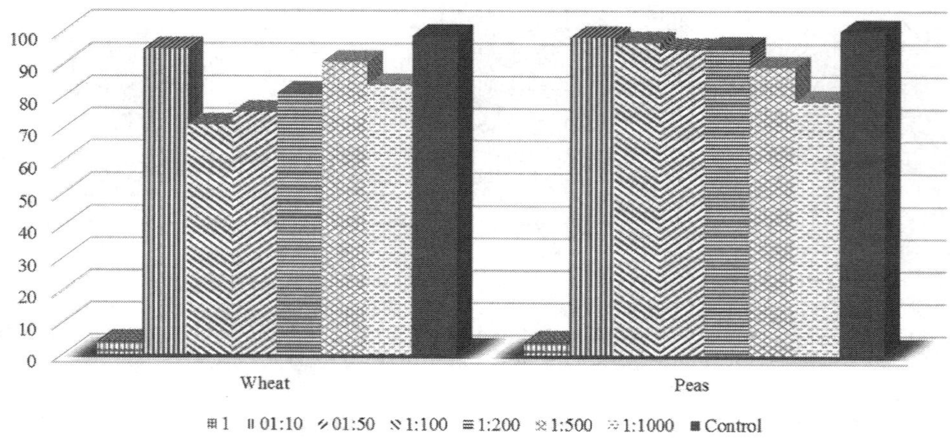

Figure 10.4 Similarity (%) of cultivated plants with the addition of processed substrate in different dilutions in the second experiment.

Therefore, an excessive amount of fertilizer can have an inhibitory effect on the germination of seeds of cultivated plants. It is necessary to investigate in greater detail the kinetics of the impact of these wastes on the growth and development of cultivated plants, especially those that can be used in biological reclamation.

10.4 CONCLUSIONS

Having conducted the analysis of the research, we can conclude that the use of cyanobacteria after anaerobic fermentation as a fertilizer or component to create a growth substrate is the most promising solution. Analytical studies confirm the absence of limiting hazardous components that would adversely affect the growth and development of plants. The presence of a significant amount of organic matter, basic nutrients, macro- and microelements, as well as fulvic and humic acids in these wastes, which are components for soil fertility reproduction and are also used in agriculture as plant growth stimulants, was determined.

REFERENCES

Cavinato, C. et al. 2017. Biogas production from microalgae. Microalgae-based biofuels and bioproducts. *Woodhead Publishing Series in Energy*: 155–182. doi:10.1016/B978-0-08-101023-5.00007-8

De Caire, G.Z. et al. 2000. Short communication: Changes in soil enzyme activities following additions of cyanobacterial biomass and exopolysaccharide. *Soil Biology and Biochemistry* 32(13): 1985–1987.

Kvaternyuk, S. et al. 2017. Increasing the accuracy of multispectral television measurements of phytoplankton parameters in aqueous media. *17th International Multidisciplinary Scientific GeoConference SGEM 2017: SGEM2017 Vienna GREEN Conference Proceedings*, Vol. 17, Issue 33, pp. 219–225., Bulgaria.

Malovanyy, M. et al. 2016. Production of renewable energy resources via complex treatment of cyanobacteria biomass. *Chemistry & Chemical Technology* 10(2): 251–254. doi:10.23939/chcht10.02.251

Malovanyy, M. et al. 2018. Two stage treatment of solid waste leachates in aerated lagoons and at municipal wastewater treatment plants. *Eastern-European Journal of Enterprise Technologies* 1(10): 23–30. doi:10.15587/1729-4061.2018.122425

Malovanyy, M. et al. 2019. Perspective technologies of the treatment of the wastewaters with high content of organic pollutants and ammoniacal nitrogen. *Journal of Ecological Engineering* 20(2): 8–15. doi:10.12911/22998993/94917

Obana, S. et al. 2007. Effect of Nostoc sp. on soil characteristics, plant growth and nutrient uptake. *Journal of Applied Phycology* 19(6): 641–646.

Osman, M.E.H. et al. 2010. Effect of two species of cyanobacteria as biofertilizers on some metabolic activities, growth, and yield of pea plant. *Biology and Fertility of Soils* 48(8): 861–875.

Pandey, K.D. et al. 2005. Cyanobacteria in alkaline soil and the effect of cyanobacteria inoculation with pyrite amendments on their reclamation. *Biology and Fertility of Soils* 41(6): 451–457.

Petruk, V. et al. 2012. The spectral polarimetric control of phytoplankton in photobioreactor of the wastewater treatment. *Proceedings of the SPIE* 8698: 86980H.

Petruk, V. et al. 2015. The method of multispectral image processing of phytoplankton for environmental control of water pollution. *Proceedings of the SPIE* 9816: 98161N.

Saadatnia, H., Riahi, H. 2009. Cyanobacteria from paddy fields in Iran as a biofertilizer in rice plants. *Plant, Soil and Environment* 55(5): 207–212.

Sakalova, H. et al. 2019. The research of ammonium concentrations in city stocks and further sedimentation of ion-exchange concentrate. *Journal of Ecological Engineering* 20(1): 158 –164. doi:10.12911/22998993/93944

Shkvirko, O., Tymchuk, I., Malovanyy, M. 2018. The use of bioindication to determine the possibility of sludge recovery after biological treatment of wastewater. *Environmental Problems* 3(4): 258–264.

Shkvirko, O., Tymchuk, I., Malovanyy, M. 2019. Adaptation of the world experience in the utilization of sewage sludge to the ecological conditions of Ukraine. *Scientific Bulletin of UNFU* 29(2): 82–87. doi:10.15421/40290216

Singh, J.S., Kumar, A., Singh, M. 2019. Cyanobacteria: A sustainable and commercial bio-resource in production of bio-fertilizer and bio-fuel from waste waters. *Environmental and Sustainability Indicators* 3–4: 100008. doi:10.1016/j.indic.2019.100008

Zelenko, Y., Malovanyy, M., Tarasova, L. 2019. Optimization of heat-and-power plants water purification. *Chemistry & Chemical Technology* 13(2): 218–223. doi:10.23939/chcht13.02.218

Chapter 11

Biomass of Excess Activated Sludge from Aeration Tanks as Renewable Raw Materials in Environmental Biotechnology

Alona V. Pasenko, Oksana V. Maznytska, and Tatyana M. Rotai
Kremenchuk Mykhailo Ostrohradskyi National University

Larysa E. Nykyforova
National University of Life and Environmental Sciences of Ukraine

Andrzej Kotyra
Lublin University of Technology

Bakhyt Yeraliyeva and Gauhar Borankulova
M. Kh. Dulaty Taraz Regional University

CONTENTS

11.1 INTRODUCTION

Active sludge is the biocenosis of wastewater treatment plants for aerobic biological wastewater treatment. It carries out the biodegradation of pollution of municipal, industrial wastewater. On the basis of this environmental biotechnology, natural processes of self-purification in the aquatic environment are employed with the mandatory condition of aeration. In the aeration tank, biochemical processes are implemented, which are part of the biological cycle of substances, including utilization, transformation and mineralization of organic substances. Aerobic fermentation of effluents with organic pollutants is carried out by a specific complex – activated sludge biomass, consisting of specific microorganisms (bacteria, fungi, protozoa, and other organisms) in the biocenosis composition, formed under artificially organized environmental conditions caused by the composition of wastewater.

DOI: 10.1201/9781003177593-11

An effective wastewater treatment process takes place under the optimal cultivation conditions: temperature, pressure, and the presence of a sufficient amount of necessary nutrients that help optimize the flow of energy metabolism and biosynthesis of the cellular substance and, as a result, an increased content of lipid fraction in the activated sludge biomass cells (Malovanyy et al. 2019). In urban ecosystems at municipal wastewater treatment plants, after mechanical and biological wastewater treatment, a significant amount of waste is generated – wastewater sludge, including excess activated sludge, which is a valuable biosubstrate. When processing every 1,000 m^3 of wastewater, up to 300 dm^3 of dehydrated excess activated sludge (80% humidity) is formed (Pasenko & Nykyforova 2018). Globally, the annual increase of activated sludge reaches several tens of millions of tons (in terms of dry matter weight). The search for a technological solution to the issue of handling these wastes is one of the important environmental problems of sustainable development of urban ecosystems.

The technology for processing sewage sludge provides for the drying of excess activated sludge at sludge sites for further storage or disposal (Pasenko & Nykyforova 2018, Yurchenko & Astapova 2010). As a biosubstrate, the biomass of excess activated sludge is in 70%–80% composed of the compounds with macroergic chemical bonds – lipids, proteins, and carbohydrates – which can be extracted and used as a renewable raw material source in industrial production cycles (Yurchenko & Astapova 2010, Pasenko 2012). Therefore, the classical scheme for the treatment of sewage sludge with their storage at silt sites is not rational and requires further improvement, namely, the search for the environmentally safe ways of disposing of these wastes.

A significant problem for specialists in the development of environmental biotechnologies for the disposal of excess activated sludge is the possible accumulation of heavy metals in the cells of activated sludge organisms, high sludge moisture, the presence of pathogenic microorganisms, and the negative effect when introduced into ecosystems on soil, groundwater and atmospheric air. When storing sediments, an unpleasant odor is observed during their decay on sludge sites. The pollutants adsorbed from wastewater and accumulated by activated sludge can migrate to soil and groundwater when sludge is stored in silt sites, polluting the natural environment and forming a number of negative environmental consequences. Therefore, at present, specialists have developed a number of schemes for the complex processing of excess activated sludge, including the removal of valuable chemical components from its cell mass for their further use in various production cycles (Pasenko et al. 2018). This can significantly reduce the anthropogenic pressure on the environment and decrease the consumption of traditional raw materials in the production of industrial products. The general scheme of complex processing of biomass of excess activated sludge is presented in Figure 11.1.

A promising product of the complex processing of excess activated sludge is bioplastics (Pasenko et al. 2018). Innovative biological technologies for generating energy, chemical compounds with macroergic bonds from renewable sources, namely, from the biomass of organisms, are an alternative to the energy-safe growth and improvement of the country's industrial sector. The polymer compounds for large tonnage chemical production of plastic are obtained from petroleum hydrocarbons or natural gas. The specified technology for the production of plastic requires large quantitative expenditures of natural combustible substances and contributes to the formation of significant volumes of greenhouse gases. The production cycle of synthetic plastic

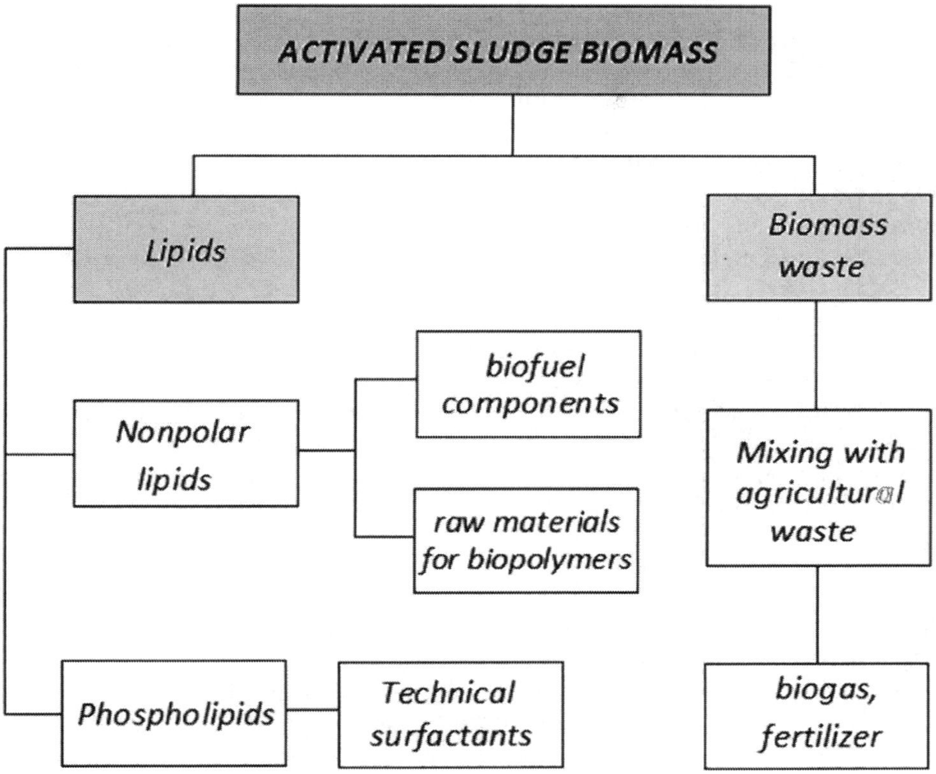

Figure 11.1 The scheme of complex processing of activated sludge biomass.

significantly affects the ecological state of the environment. These environmental impacts are significantly lower in the production of biopolymers (bioplastics).

It is known that, for example, one plastic bag takes about 15 years to biodegrade, plastic containers – about 300 years, at the same time, decomposition of bioplastics will take only 1–2 years (Pasenko & Nykyforova 2018). Thus, it is the most appropriate environmental-friendly technology for the production of bioplastics and, above all, from renewable biosubstrate – spent biomass of activated sludge organisms. In general, in an environment in which all components are interconnected, any environmental problem passes into another, and this, in turn, forms a whole complex of ecological disturbances in the ecosystem, which directly or indirectly affects each organism of a certain ecotope. By reducing the negative impact on the state of the environment, it is possible to maximize the normalization of the environmental indicators of the territory of not only a single country but the entire region (Pasenko 2012. Ukraine has significant potential for replacing non-renewable energy sources, raw materials (oil, gas) in the production cycles of various products with renewable sources, which is extremely relevant from an economic and environmental point of view.

It should be noted that in the world practice, corn and potato starch are most often used as a substrate for the production of bioplastics. A promising area is the use of organic waste as a substrate for the production of bioplastics. According to the results of the analysis of previous developments (Malovanyy et al. (2019a), bioplastics are recommended to be produced from the lipids obtained by the extraction of organisms from the spent biomass. The direction regarding the use of biomass of excess activated sludge as a naturally renewable raw material for bioplastics has been studied too little. The basic chemical component of sludge as a raw material in bioplastics technology is the lipid fraction of excess activated sludge cells. Under favorable environmental conditions in treatment plants, activated sludge microorganisms produce mainly polar lipids (e.g. phospholipids). The phospholipids extracted from biomass are subjected to purification and are subsequently widely used as technical surfactants and in cosmetology.

It is not advisable to consider the use of phospholipids as food and biologically active additives in view of the possible content of heavy metals in them. Polar lipids are the structural components of all living cells, are part of the cytoplasmic, mitochondrial, and other membranes, play a significant role in membrane permeability, and are responsible for the location of respiratory chain enzymes and electron transfer. Under unfavorable conditions restricting the growth of activated sludge bacteria (stressful conditions), neutral lipids accumulate in the cytoplasm of cells in the form of lipid drops, which are the main spare components of the cell (Gouveia 2011). Non-polar lipids isolated from cells are widely used in the production of biofuels, biopolymers, and other products. The residues of biomass (proteins, carbohydrates) after extraction of lipids are purified and used to obtain biogas, liquid, and solid fertilizers (Shvets 2009, Malovanyy et al. 2019b).

The results of the isolation of the lipid fraction from the biomass of activated sludge are largely dependent on the reagent during extraction and the physiological state of the cells of excess active silt. The lipids thus obtained from activated sludge can serve as an alternative raw material when replacing synthetic plastic technology with bioplastics. The proposed technology for the production of bioplastics from excess activated sludge is under development. Therefore, the aim of this work is to study activated sludge as a substrate of environmental biotechnology and extract lipids from biomass of excess activated sludge to obtain a lipid fraction for further use in the production of bioplastics.

11.2 MATERIALS AND METHODS

Excess activated sludge is proposed for use in the bioplastics production technology as a raw material, a substrate for the extraction of lipids, which has certain advantages. Firstly, the quantitative and qualitative composition of the biomass of excess activated sludge indicates the specific functioning conditions of the aeration tank ecosystem. The composition of the activated sludge biocenosis includes a large number of representatives of microflora and microfauna: filamentous bacteria, lower fungi, yeast, flagellates, sarcodes, ciliates, rotifers, waterworms, and in small quantities other multicellular invertebrates (water mites, gastrotrichs, etc.) (Pasenko & Nykyforova 2018). Activated sludge consists in 95% of prokaryotes, whose cells contain inclusions that act as a reserve substance, including, for example, lipids (Pasenko et al. 2018).

Secondly, the advantage of this substrate (biomass) when used in environmental biotechnology is its high productivity when operating in a wastewater treatment plant. The energy potential of activated sludge biomass is much higher than any crop. The metabolic activity of activated sludge is due to the ability of microorganisms – representatives of the *Azotobacter*, *Bacillus*, *Bacterium*, *Pseudomonas* families, etc. to actively use organic substances for nutrition, namely, the available forms of nitrogen, phosphorus, and potassium of various compounds contained in wastewater. In the process of nutrition, microorganisms receive organic substances as "building blocks" for building their body, resulting in an increase in the activated sludge biomass (Yurchenko & Astapova 2010).

The representatives of excess activated sludge from aeration tanks are characterized by a high efficiency coefficient (COP) of the consumption of organic pollutants during the biological treatment of wastewater. Along with this, there is practically no loss of biomass. The increase in biomass is directly proportional to a specific oxidation rate of the substrate. Wastewater treatment in aeration tanks occurs under optimal conditions: temperature up to 25°C, certain pressure, and high content of nutrients contribute to the rapid growth of biomass and the accumulation of spare substances by microorganisms, including lipids. The total content of lipids, namely, phospholipids, in biomass cells of excess activated sludge ranges from 20% to 30% by weight of dry matter. The quantity and qualitative composition of the lipid yield upon extraction from cells largely depends on the conditions of the technological process of microorganism cultivation, including the functioning and productivity of activated sludge in aeration tank. An important issue determining the yield of lipids is the species composition of the activated sludge biocenosis that forms in wastewater treatment plants under the given technological conditions. The renewable bioenergy raw materials synthesized by the cells of microorganisms – lipids – using standard biotechnological processes can be further processed into bioplastics (Pasenko & Nykyforova 2018, Pasenko et al. 2018).

The features of the biocenosis composition of activated sludge determine the chemical composition of its biomass, including the number and composition of the lipid fraction of cells. Therefore, to study the species composition of organisms of excess activated sludge, a hydrobiological analysis of the sludge of treatment facilities was conducted in the work.

Extraction of the lipid fraction from a raw organic substrate is an important method in the implementation of environmental biotechnologies. The technology for the integrated use of biomass of excess activated sludge involves the following stages: extraction of biomass of sludge from a suspension of sewage sludge; disintegration of the cell walls of microorganisms for maximum extraction of the lipid fraction. Lipids are extracted with a mixture of polar and non-polar solvents, while phospholipids will be contained in the fraction of the polar solvent, and triglycerides will be contained in the non-polar fraction. The scheme for producing lipids on the example of phospholipids during extraction from the biomass of microorganisms is presented in Figure 11.2.

According to the scheme (Figure 11.2), the biomass is concentrated as an organic substrate; then, the biomass cells are destroyed for more efficient extraction of lipids. Extraction is carried out with a mixture of polar and non-polar extractants for 2.5–3 h. The extractant is distilled off, and the lipid mixture is divided into fractions: the fraction containing non-polar lipids is used in further production. After the reaction with methanol, nonpolar lipids form fatty acid methyl esters and technical glycerin in the presence of a catalyst, which is used in many industries (detergents and cosmetics, agriculture, textiles, paper, and leather). The completeness of lipid

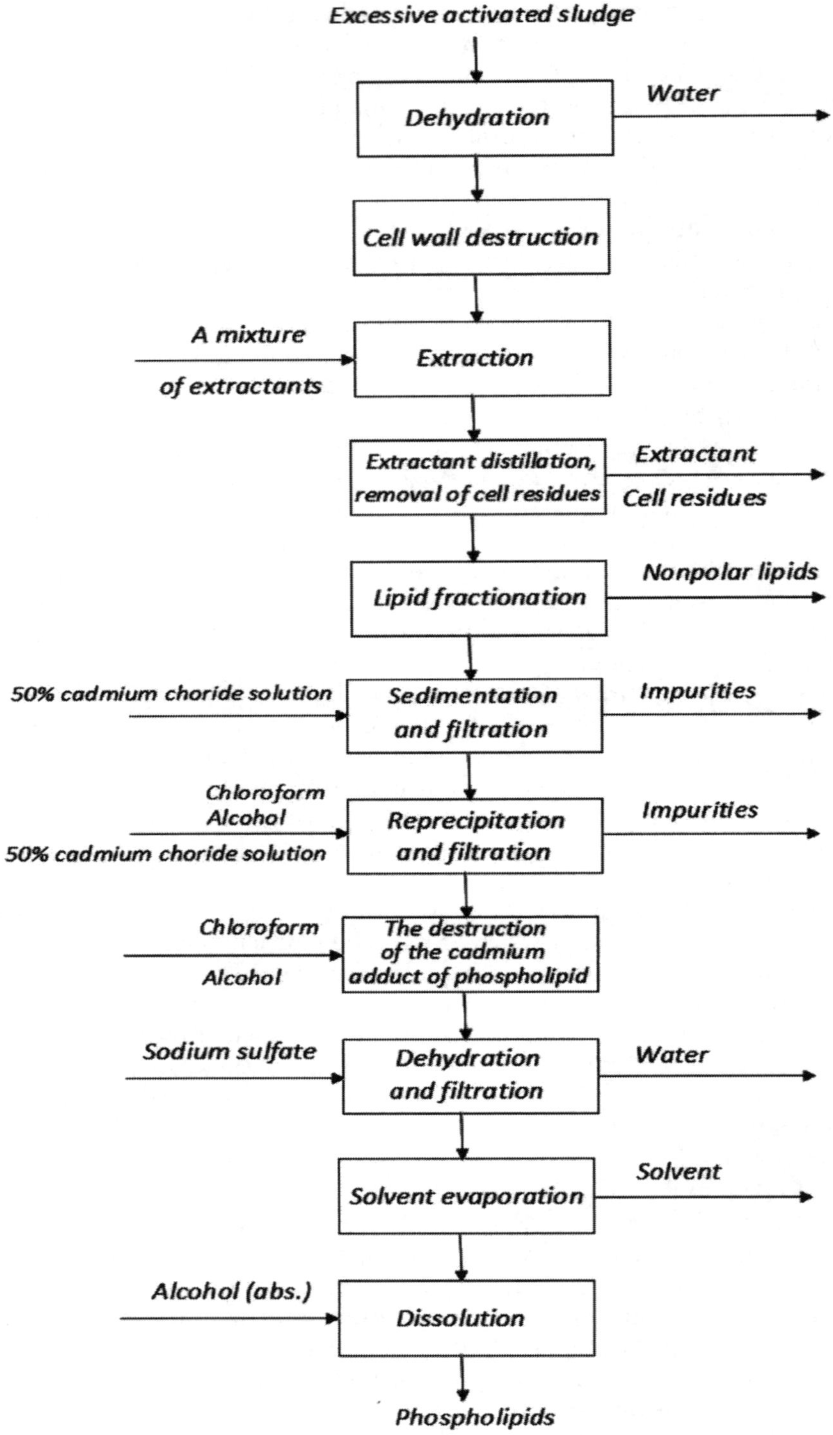

Figure 11.2 The scheme of the extraction of phospholipids from an organic substrate.

extraction is greatly influenced by shredding of biomass (the amount of extracted lipids after cell wall destruction can be 1.1–2 times higher compared to extraction from non-destroyed biomass cells) (Nykyforov et al. 2019).

Bacterial cells can be destroyed by freezing three times in liquid nitrogen, ultrasound, using a ball mill or French press, although not all of these methods are convenient for preparing large volumes of extract. In the case of small volumes of the test material, the cells can be triturated with alumina. The extraction of lipids from the activated sludge biomass is a process that is characterized by high energy costs, which greatly complicates the use of biomass as a raw material for the production of a wide range of products on an industrial scale (Chianell et al. 2013, Nagornov 2010). The extraction of lipid of activated sludge biomass can be divided into extraction from wet biomass (in which the extraction of the target product is difficult by the presence of culture fluid, which leads to a low yield of the product during the extraction process), as well as the extraction of lipids from dry activated sludge biomass, the disadvantage of which is high energy consumption for drying biomass.

The extraction method using a mixture of chloroform: methanol (1:1 vol./vol.) was developed in 1951 by Folch and others and is fast and effective (Halim 2012, Malovanyy et al. 2016). A mixture with a ratio of methanol and chloroform of 2:1 (vol./vol.) is also used – the method of Bly and Dyer. However, due to the high toxicity of chloroform, its use is undesirable. As a substitute for chloroform, dichloromethane is used in a ratio of methanol: dichloromethane 1:2 (vol./vol.) – Chen's method. A solvent mixture with low toxicity hexane: isopropanol 3:2 (vol./vol.) is more preferable as a substitute for a mixture of chloroform: methanol. After two-phase separation, the light organic phase (hexane with some isopropanol) contains most of the lipids (neutral and polar), and the lower aqueous phase (water with some isopropanol) contains proteins and carbohydrates. Pure alcohol (e.g. butanol, isopropanol, and ethanol) is a cheap volatile extractant and is able to form hydrogen bonds with the protein-lipid membrane complexes due to its polar nature.

However, its polar nature is also a drawback, since it limits the interaction with autonomous globules of neutral lipids. For this reason, when used as a solvent, in the extraction of activated sludge lipids, alcohol is almost always used in combination with a non-polar organic solvent, such as hexane or chloroform, to provide general extraction of both forms of neutral lipids (both autonomous beads and membrane-associated complexes). Soxhlet's extraction is a method that enables to completely extract all the lipids present in the cells of microorganisms of activated sludge, resulting in complete recovery. Due to the circulation of the solvent in the extractor due to evaporation and condensation, the biomass cells interact with a fresh organic solvent, as a result of which the greatest driving force is maintained, while the consumption of solvent is minimized.

During the experimental extraction of lipids from the biomass of excess activated sludge, the method of extraction of neutral and total lipids using non-polar solvents was used. Hexane, having less toxicity than other solvents used for extraction, for example, chloroform and diethyl ether, was used as an extractant. However, one must also take into account that this method leads to the loss of a large number of polar lipids (phospholipids).

11.3 STUDY OF THE BIOCENOSIS COMPOSITION OF ACTIVE SLUDGE

Hydrobiological analysis of the biocenosis composition of activated sludge was carried out on the basis of the laboratory of treatment facilities in the city of Kremenchuk. The wastewater bio-treatment process at the WWTP facilities is carried out in the aeration tank. As a result of the technological reconstruction of the facilities, they were rebuilt, and the aerotank at the time of the experimental research worked as a bioreactor with aerobic-anaerobic zones in the corridors. As a result, the species adapted to the anoxic conditions of existence predominated in the biocenosis of activated sludge. In the biocenosis of activated sludge, its indicator organisms were studied, the presence of which, the features of vital functions, development and reproduction indirectly testify to the specificity and effectiveness of wastewater treatment processes in the aeration tank. The moisture content of the initial biosubstrate: activated sludge, the average value of which was 99.52%, and the ash content, the average value of which was 42.2%, was determined in the work.

The biocenosis of activated sludge is formed from microorganisms that are the most resistant to wastewater pollution with appropriate nutritional needs, and the species diversity, for example, protozoa, is determined by the degree of decomposition of organic pollutants during bio-treatment of effluents in the aeration tank. The nature of the reaction of activated sludge biocenosis to the adverse effects of environmental factors is manifested in a decrease in species diversity with a maximum number of the most stable species. The important factors that negatively affect the formation of activated sludge are the imbalance in the ratio of nutrients in wastewater, fluctuations in pH, the influence of light, temperature, aeration, salinity, etc.

Experimental studies on the composition of activated sludge in wastewater samples using hydrobiological analysis revealed many small, colorless flagellates of the Bodo (*Sarcomastigofora*, *Zoomastigoforae*) genus. The identification of a large number of colorless flagellate Mastigofora in the biocenosis of the bioreactor indicates sludge overload on organic matter, nitrogen compounds, and oxygen deficiency, which is typical for the anoxic conditions. A large number of tortoise amoebas of the *Sarcodina* class (*Sarcomastigofora*) with disk-shaped turtles *Arcella discoides* and *A. vulgaris* were observed. Many cysts of tortoise rhizopods, formed due to adverse conditions (lowering the temperature, lack of food, etc.), were found.

In the biocenosis of fouling on carriers, many ciliates (*Ciliophora*) were observed. In treatment facilities, ciliates, after bacteria, make up the largest group of biocenosis organisms. Out of the ciliates, microorganisms of the *Kinetophragminofora* class, the most primitively organized organisms with a large body, were often found. Equidimensional ciliates of the *Oligohymenofora*, free-floating forms, were also observed. Tailed ciliates or ciliates – *Paramecium caudatum*, have a large body, similar to a spindle, elongated, round in cross-section, covered with cilia, longer on the back. They are adapted to exist under the conditions of low concentration of dissolved oxygen in water, characteristic of anoxide bioreactors.

A lot of ciliates were found in the fouling of the gastrociliary infusoria of the *Polyhymenophora*: *Aspidisca costata, A. turrida*, *Stylonyhia pustulata*, *Euplotes sharon*. These ciliates settle in biofouling. There were many attached round ciliate infusoria (*Peritriha*): *Vorticella microstoma* and *V. submicrostoma* – single, small forms attached to the detritus particles using a stem. These species develop in an environment highly polluted with organic substances, which can be traced in the anoxide bioreactor. In the process of purifying water, they play the role of sedimentators.

The colonial forms of peritrichs: *Epistylis plicatilis* and *Opercularia glomerata* occurred to a lesser extent. Occasionally, *Didinium* predatory ciliates that feed on moths were observed. In fouling, predatory sucking ciliates (*Suctoria*) were observed: *Podophrya, Tokophrya*. In the biocenosis of biofouling in the anoxide bioreactor, rotifers (*Nemathelminthes, Rotifera*) are quite often found – primary-striped worms that exist even at low oxygen concentrations and are resistant to the content of hydrogen sulfide or methane. They contribute to the purification of water and the mineralization of biomass fouling. The primary streaky organisms were free-living roundworms *Nematoda* (*Nemathelminthes*). In the fouling, few *Oligoschaeta* worms (*Annelida*) were observed. They settled in fouling and can withstand a large oxygen deficiency.

The absence of small ciliates Vorticella microstoma and V. submicrostoma, characteristic of the anoxide bioreactor in the aerobic section of the bioreactor, should be noted. Moreover, the colonial forms of *Epistylis plicatilis, Opercularia* and predatory sucking ciliates were only in small quantities. A large number of different types of rotifers were found in the biocenosis of biofouling: *Philodina roseola, Cathypna luna*, and *Rotaria rotatoria*, in contrast to anoxide bioreactors, where there were few of them (found in the anaerobic bioreactor), because rotifers are sensitive to a lack of dissolved oxygen in water. They lose mobility, stretch out and gradually die off.

Oligoschaeta – Tubifex tubifex worms were found on carriers. The presence of rotifers and worms in the aerobic bioreactor, the organisms of the highest links of the trophic chains of the hydrobiocenosis of the treatment plant, provides: improvement of the water purification process, since these organisms eat detritus, bacteria, and protozoa, preventing their removal from the bioreactor along with purified water, and a decrease in the growth of biomass of microorganisms. As a result, the processing and disposal costs of excess activated sludge are reduced; mineralization of biomass occurs, which improves the sedimentation properties of sediment.

11.4 STUDY OF THE LIPID COMPOSITION OF EXCESS ACTIVATED SLUDGE BIOMASS

For the implementation of environmental biotechnology for the production of bioplastics, an important step is to study the chemical composition of activated sludge as a raw material substrate for the technology. The key material base of the technology is the lipid fraction extracted from the cells of the organisms of the biocenosis of activated sludge aeration tanks. Active sludge contains both the prokaryotic and eukaryotic organisms.

Lipids are contained in the plasma membranes of all living cells. The content of components in the cell membrane of most prokaryotes is distributed as follows: proteins – up to 50%, lipids – up to 30%, carbohydrates – up to 20% (calculated on dry weight), for eukaryotes, these figures are respectively: proteins – up to 70%–80% lipids up to 15%–25%, carbohydrates up to 5%–15%. However, these values can vary greatly and lipids can account for up to 50% of the dry weight of the cell membrane. The main lipid component of the cell membrane of activated sludge bioagents is phospholipids – derivatives of 3-phosphoglycerol. It is widely represented in the cell membranes and various glycolipids.

Bacteria contain a large number of different lipids, including sphingolipids and neutral lipids. The vast majority of prokaryotes lack sterols. The composition of prokaryotic membrane lipids contains acids, which, as a rule, are absent in the membranes of eukaryotic cells. These acids include cyclopropane fatty acids containing one or more three-membered rings attached along a hydrocarbon chain. Rarely, there are branched fatty acids with a 15–17 carbon chain, which mainly occur in prokaryotes.

The basis of the lipid fraction of activated sludge cells is saturated or monounsaturated fatty acids. Polyunsaturated fatty acids of the lipid fraction are absent in prokaryotic cells (Pasenko & Nykyforova 2018, Pasenko et al. 2018). Lipids accumulate in the cells in the form of granules, sharply refract light and are, therefore, clearly visible in a light microscope. A reserve substance of this kind is a polymer of β-hydroxybutyric acid, which accumulates in the cells of many prokaryotes. In some hydrocarbon oxidizing bacteria, poly-β-hydroxybutyric acid makes up to 70% of the dry matter of the cells. Lipid deposition in the cell wall of activated sludge organisms occurs under the conditions when the medium – wastewater is saturated with carbon compounds and in deficiency – nitrogen compounds. Lipids for hydrobiont cells serve as a source of carbon and energy.

In order to determine the total lipid content in the activated sludge biomass, it was dried to an air-dry state and triturated with alumina (cell walls were destroyed due to the presence of microroughnesses on the abrasive particles). The crushed biomass for extracting the lipid fraction was transferred to a separatory funnel and poured with a mixture for extraction of hexane: water in a ratio of 1:1 (vol./vol.) in an amount of $100\,cm^3$.

After vigorous shaking for 20 minutes, the mixture was left to settle (two layers formed). The lower layer (polar water, intensely colored, brown) contained the undissolved remains of activated sludge. The upper part of the funnel contained a non-polar, less intensely colored liquid-phase mixture of hexane with extracted lipids. The aqueous layer was drained, and the upper (hexane + lipids) was transferred to a porcelain bowl and evaporated in a water bath. The amount of extracted lipids was determined gravimetrically. The lipid content was 2.35% of the dry biomass of activated sludge.

The fatty acids in bacterial cells are predominantly part of acyl-containing lipid molecules. The chain length of fatty acids is 10–20 carbon atoms. In this case, the acids with 16 and 18 carbon atoms prevail. The content of fatty acids (%) in activated sludge is shown in Figure 11.3.

As can be seen from Figure 11.3, palmitic acid (C16:0) and stearic acid (C18:0) are contained from saturated acids, and palmitoleic acid (C16:1), oleic acid (C18:1), and linoleic acid (C18:2) are present from unsaturated acids. The obtained lipid fraction was studied by thin layer chromatography (plates with a thin layer of Sorbfil silica gel). In order to increase the sorption ability of silica gel and remove moisture, the plates were activated by keeping in a thermostat at a temperature of 100°C–110°C for 30 minutes. The studied lipid mixture in an amount of $0.1\,cm^3$ was applied to the plates in the form of dots with a diameter of 0.2 cm at a distance of 1.5 cm from the edge of the plate and 1.0 cm from each other.

The plates with deposited samples were placed in a chromatographic chamber with a solvent layer of 0.7–1.0 cm. For the separation of neutral lipids, a mixture of hexane: diethyl ether: glacial acetic acid in the ratio 73:25:2 (experiment #1) was used.

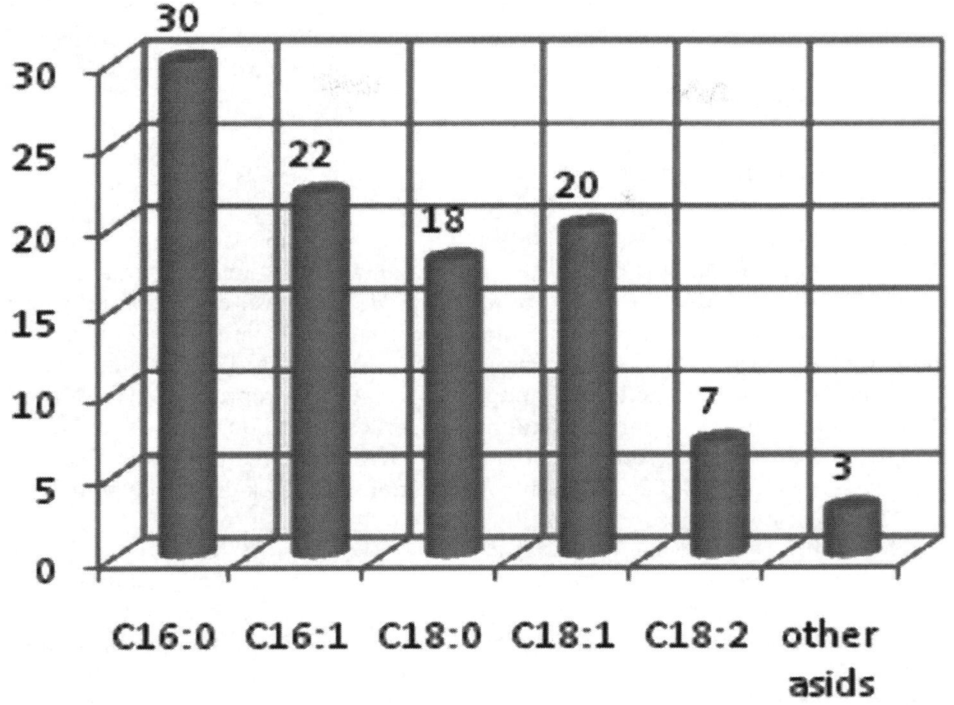

Figure 11.3 The content of fatty acids (%) in activated sludge.

A mixture of chloroform: methanol: water was used for the separation of phospholipids in a ratio of 65:25:4 (experiment #2). Lipid distillation continued until the solvent front reached a level of 0.5–1.0 cm distant from the top of the plate. As a control, a standard chloroform-methanol mixture of lipids consisting of lecithin, fatty acids (palmitic, stearic, oleic) was used, triacylglycerol.

As a developer of the chromatogram in experiment #1, iodine pairs (non-destructive reagent) were used. After drying, the plate was placed in a desiccator with crystalline iodine for 5–10 minutes. Lipids appeared on the chromatogram as yellow-brown spots. For experiment #2, a chromatogram developer, phosphonomolybdenum acid, was used (the plate was irrigated with a 5% solution of phosphoromolybdenum acid in ethanol, heated to a temperature of 80°C–90°C). Lipids were stained dark blue against a yellow-green background. The lipid separation results were recorded in Figure 11.4.

After fractionation of lipids in the solvent system hexane: diethyl ether: glacial acetic acid in a ratio of 73:25:2 (experiment #1), chromatograms were obtained on which lipid fractions were concentrated in the following sequence, starting from the start line: phospho and glycolipids. Monoacylglycerols and diacylglycerols, free fatty acids, triacylglycerols. Hydrocarbons are adjacent directly to the front line of the solvent.

In the solvent system chloroform: methanol: water in a ratio of 65:25:4 (experiment #2), starting from the start line, lipids were distributed in the following sequence: phosphatidylcholine; phosphatidylethanolamine; phosphatidylglycerol; and diphosphatidylglycerol. Neutral lipids adjoin the front line of the solvent (finish). From Figure 11.4, it can be seen that the main of the isolated phospholipids are phosphatidylethanolamine, phosphatidylcholine, phosphatidyl glycerol, and traces of diphosphatidyl glycerol, which is due to the presence of activated sludge protozoa and worms in the biocenosis.

11.5 CONCLUSIONS

The use of biomass of excess activated sludge as a renewable organic raw material in ecological biotechnology of bioplastics production was substantiated, and the feasibility of using the proposed waste as a source of lipid fraction was experimentally proven.

According to the results of hydrobiological analysis, the presence of numerous indicator microorganisms of the genera Arcella, Aspidisca, Paramecium, Euplotes, Vorticella and others in activated sludge was found. The representatives of the Epistylis, Opercularia, Podophrya, Tokophrya were less common. Many small colorless flagellates of the genus Bodo, Mastigofora genera were identified, which indicates sludge overload on organic matter, nitrogen compounds, lack of oxygen, which is typical for anoxic conditions.

Thus, the calculated technological expectations of the formation of aerobic-anaerobic conditions in different aeration tank corridors are empirically confirmed as a result of the reconstruction of treatment facilities. In the biocenosis of the anoxide bioreactor, rotifers (Rotifera) which are the organisms on the highest level in the trophic chain of hydrobiocenosis, regulate the number of bacteria and protozoa. These organisms are quite abundant in the environment with low oxygen concentrations and

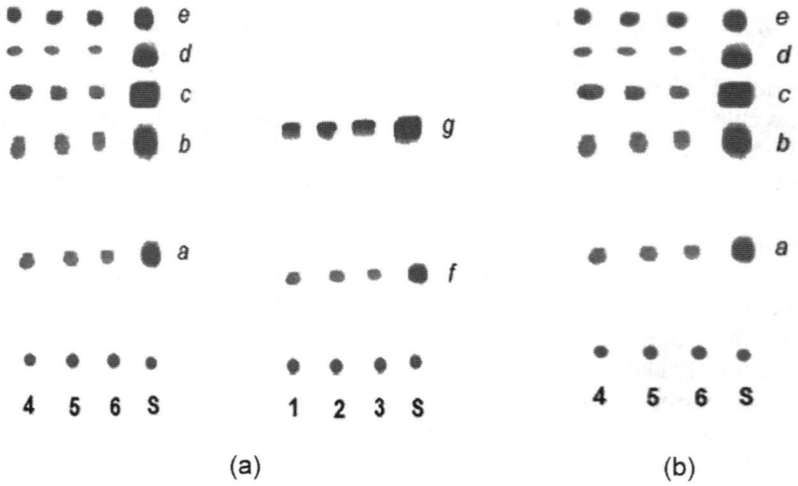

Figure 11.4 Thin layer chromatography of activated sludge lipids: (A) – neutral lipids (experiment # 1): S – standard; 1, 2, 3 – lipid samples; (B) – polar lipids (phospholipids, experiment #2): S – standard; 4, 5, 6 – lipid samples; a – phosphatidylcholine; b – phosphatidylethanolamine; c – phosphatidylglycerol; d – diphosphatidylglycerol; e – neutral lipids; f – saturated fatty acids; g – triacylglycerol.

high content of hydrogen sulfide and methane. Their presence in the activated sludge of biological facilities reduces the cost of processing and disposal of excess activated sludge; mineralization of biomass, which improves the sedimentation properties of sludge with further dehydration.

An analysis of the data obtained in the course of the research on the lipid composition of the sludge biomass of aeration tanks confirms the possibility of using excess activated sludge as a renewable raw material for the production of lipids in ecological biotechnology of bioplastics production. The implementation of this process enables to comply with the environmental requirements regarding energy and resource conservation, and also solves the problem of disposal of excess activated sludge. At the same time, firstly, to obtain lipids, free municipal waste was used as raw material; secondly, it is possible to obtain a lipid extract for its further use in various biotechnological or chemical processes (production of bioplastics or biofuels, biopolymers, surfactants, and other products).

REFERENCES

Chianelli, R., Hildebrand, C. & Rodriguez, J. 2013. *Process for separating lipids from a biomass.* Patent No. 8476060 US 12/758,480.

Gouveia, L. 2011. *Microalgae as a Feedstock for biofuels.* Berlin: Springer.

Halim, R. 2012. Extraction of oil from microalgae for biodiesel production: a review. *Biotechnology Advances* 30: 710–731.

Malovanyy, M., Moroz, O., Hnatush S. et al. 2019a. Perspective technologies of the treatment of the wastewaters with high content of organic pollutants and ammoniacal nitrogen. *Journal of Ecological Engineering* 20(2): 8–15.

Malovanyy, M., Zhuk, V., Nykyforov, V., Bordun, I., Balandiukh, I. & Leskiv, G. 2019b. Experimental investigation of Microcystis aeruginosa cyanobacteria thickening to obtain a biomass for the energy production. *Journal of Water and Land Development* 43(1): 113–119.

Malovanyy, M., Nikiforov, V., Kharlamova, O. & Synelnikov, O. 2016. Production of renewable energy resources via complex treatment of cyanobacteria biomass. *Chemistry & Chemical Technology* 10(2): 251–254.

Nagornov, S.A. 2010. *Sposob izvlecheniya lipidov iz biomassy.* Patent No. 2388812 (13) C1 Russia, MPC C12N1/12 (2006.01), C12P7/64 (2006.01).

Nykyforov, V., Malovanyy, M., Aftanaziv, I., Shevchuk, L. & Strutynska L., 2019. Developing a technology for treating blue-green algae biomass using vibro-resonance cavitators. *Naukovyi Visnyk Natsionalnoho Hirnychoho Universytetu* 6: 181–188.

Pasenko, A.V. 2012. Ekologicheskii aspekt skhem obrashcheniya s otkhodami vodoochistki teploelektrostantsii. *Ekologichna bezpeka* 2(14): 29–32.

Pasenko, A.V. & Nykyforova, E.A. 2018. Biotechnology of preparationy and the use of sewage sediment as a fertilizer-meliorant. In Sobczuk, H. et al. (ed.) *Water Supply and Wastewater Disposal*: 231–241. Lublin: Wydawnictwo Politechniki Lubelskiej.

Pasenko, A.V., Pismennikova, T.S., Karlik, O.I. et al. 2018. Vyluchennya lipidiv z biomasi aktivnogo mulu dlya virobnitstva bioplastika. *Visnik KrNU imeni Mikhaila Ostrogradskogo* 5(112): 115–120.

Shvets, V.I. 2009. Fosfolipidy v biotekhnologiyakh. *Vestnik MITKhT* 2009: 4–25.

Yurchenko, V.A. & Astapova, A.V. 2010. Vyiavlenie faktorov upravleniya sedimentatsionnymi svoistvami aktivnogo ila. *Sbornik nauchnykh trudov KhNADU* 48: 12–17.

Chapter 12

The Use of Activated Sludge Biomass for Cleaning of Wastewater from Dairy Enterprises

Anatoliy I. Svjatenko, Olga V. Novokhatko, Alona V. Pasenko, Oksana V. Maznytska, and Tatyana M. Rotai
Kremenchuk Mykhailo Ostrohradskyi National University

Larysa E. Nykyforova
National University of Life and Environmental Sciences of Ukraine

Konrad Gromaszek
Lublin University of Technology

Almagul Bizhanova and Aidana Kalabayeva
Kazakh Academy of Logistics and Transport

CONTENTS

12.1 INTRODUCTION

The food industry, sufficiently developed in Ukraine, causes the emergence and accumulation of environmental problems in this sector of the economy. One of them is the generation of huge volumes of industrial wastewater contaminated with various organic and inorganic substances. Disposal of wastewater is one of the most important problems facing dairy enterprises. One of the effective methods for solving this problem is to thicken the liquid effluents by evaporating water from them. The wastewater

DOI: 10.1201/9781003177593-12

from the dairy industry is highly concentrated and difficult to oxidize. Various methods are used for their effective purification – anaerobic, aerobic, and physicochemical (Kezlya et al., 2011; Pashkov, 2011). The physical methods of wastewater treatment of food enterprises that process milk, which compete in their effectiveness with the physicochemical and biological methods, include magnetic separation (Shinkarenko et al., 2010; Zagirnyak, 1996; Zagirnyak et al., 2011).

In the process of complex milk processing with the manufacture of various products, wastewater with a high content of organic substances is formed. For the purification of these waters, the anaerobic processes yielding a valuable energy carrier – methane – are the most economically and environmentally acceptable. Anaerobic treatment compared with aerobic has several advantages: less energy is consumed, and about ten times lesser growth in biomass is achieved. This leads to lower costs for the treatment of excess sludge, which also does not need to be stabilized; the concentration of anaerobic biomass is limited only by its rheological properties. Anaerobic reactors are resistant to long interruptions in the supply of wastewater, which allows them to be used effectively for the treatment of seasonal effluents.

However, in the anaerobic systems, the oxidation rate is much lower than in the aerobic systems. This is due to the low growth rate of methanogens. Therefore, the work of modern anaerobic reactors is based on the principle of maintaining a mixture of biomass and wastewater in the structure, due to which the cleaning process is significantly intensified. This is facilitated by large doses of microorganisms (Epoyan et al., 2010).

During anaerobic treatment of concentrated wastewater, various types of treatment facilities are used, including contact digesters. The efficiency of the anaerobic process is assessed by the degree of purification, load, duration of stay of effluents in the reactor, temperature, and the volumetric rate of biogas output.

For wastewater treatment, it is possible to use digesters with mesophilic or thermophilic conditions. Thus, at 53°C, the cleaning process can be worse than at 35°C. Anaerobic purification is carried out using a complex biocenosis of bacteria in several stages (from a biochemical point of view). The bacteria of the first stage break down complex organic substances to simpler ones (organic acids, alcohols, etc.), while the bacteria of the second stage turn these substances into methane.

12.2 MATERIAL AND RESEARCH RESULTS

12.2.1 The Substrate-Microbial Complex of Hydrolysis, Acidogenesis and Methanogenesis

It was proven that the bioconversion of organic substances during methanogenesis is carried out as a multi-stage process. In this process, carbon bonds are gradually destroyed under the action of enzymes of numerous groups of microorganisms and their complexes (Pastorek et al., 2006). In modern opinion, the anaerobic conversion of any complex organic matter into biogas goes through at least four successive stages (Figure 12.1):

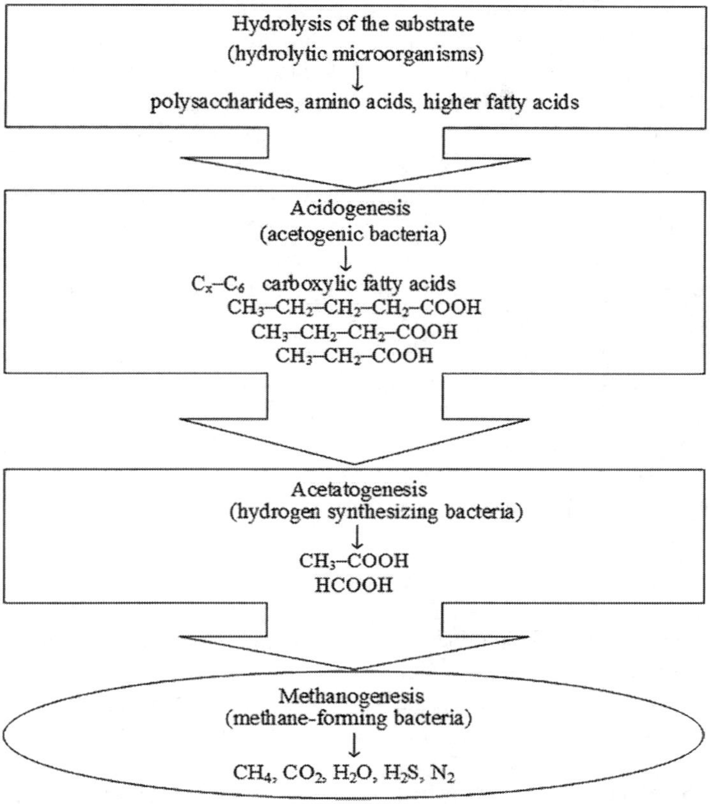

Figure 12.1 Stages of methanogenesis and intermediates (Kuris, 2016).

- the stage of hydrolysis of complex biopolymer molecules (proteins, lipids, polysaccharides, etc.) into simpler monomers (amino acids, monosaccharides, higher fatty acids, etc.);
- the stage of fermentation of the formed monomers into even simpler substances – lower acids and alcohols, while carbon dioxide and hydrogen are formed;
- the acetogenic stage, at which the direct precursors of methane are formed – acetate, hydrogen, carbon dioxide;
- methanogenic stage, which leads to the final product – methane.

At the first stage of the process, the microorganisms that exhibit cellulolytic, proteolytic, lipolytic, sulforeducing, denitrifying, and other types of activity are involved. The composition of the dominant microflora of this phase depends on the microflora composition of the incoming raw materials, as well as on the chemical nature of the intermediate decomposition products of organic substances. The number of aerobic and facultative anaerobic microorganisms in this phase reaches 10^6 cells/cm^3, and the content of obligate anaerobes is two to three orders of magnitude higher.

Among cellulose-destroying bacteria, strains of the species *Bacterioides ruminicola*, *Butyrivibrio fibriosolvens* (destroy cellulose) were found.

Among the proteolytic bacteria stand out the strains of the *Clostridium, Peptococcus, Bacterioides, Eubacterium*, and *Bifidobacterium* genera. The total number of proteolytic bacteria reaches 10^5 cells/cm^3. In this phase, 76% of organic substances are converted to higher fatty acids, up to 20% – to acetate and 4% – to hydrogen.

At the first stage of anaerobic digestion, enzymatic hydrolytic cleavage of organic substances occurs with a wide range of enzymes of the class of hydrolases secreted into the medium by anaerobic hydrolytic bacteria. Under their action, high molecular weight compounds (carbohydrates, fats, protein substances) are transformed into low molecular weight ones. In general terms, the chemistry of this step is shown in Figure 12.2 and is described by equations 12.1–12.3.

In general, the hydrolytic stage of the biomethanogenesis process can be described by the first order chemical kinetics equation, according to which the reaction rate is calculated in proportion to the current concentration of the decomposed substrate. The growth of the biomass of microorganisms is described by Mono's equation:

$$\frac{dx}{dt} = \frac{\mu_m \cdot S}{k_s + S} x, \tag{12.1}$$

where

μ_m – maximum rate of growth rate of microorganisms under given conditions;
S – the concentration of the substrate;
k_s – a constant, numerically equal to the concentration of the substrate at which the rate of growth of the culture is half the maximum.

This form of Mono's equation is similar to Michaelis-Menten's formula from enzymatic kinetics. The rate of biomass increase depends on the speed of processing of the

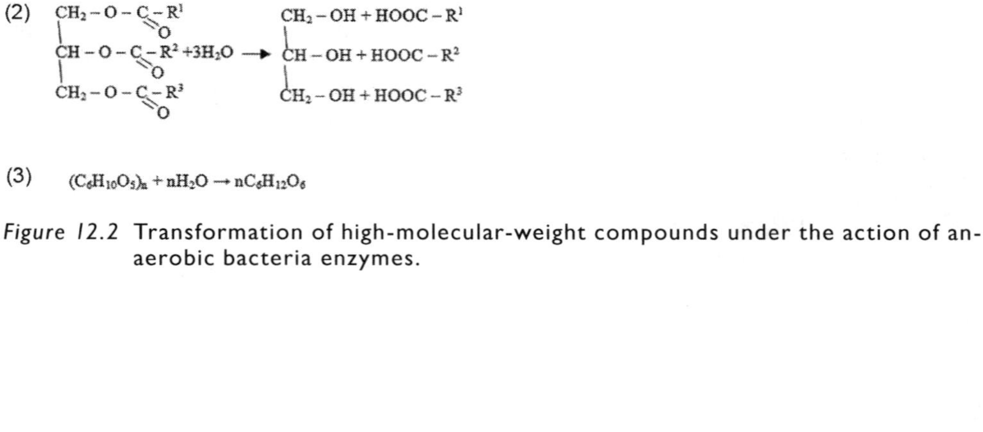

Figure 12.2 Transformation of high-molecular-weight compounds under the action of anaerobic bacteria enzymes.

substrate (limiting factor) by the enzyme in the metabolic chain. If the concentration of the enzyme per unit of biomass is designated as E_0, then according to the Michaelis law, the rate of processing of the substrate by a unit of microbial biomass has the form:

$$\frac{1}{x}\cdot\frac{dS}{dt}=\frac{k\cdot E_0\cdot S}{K_m+S}, \tag{12.2}$$

where

K_m – Michaelis constant;
k – the reaction rate constant;
E_0x – amount of total biomass enzyme x.

Thus, the total rate of decrease in the amount of substrate is equal to:

$$\frac{dS}{dt}=\frac{k\cdot E_0\cdot S\cdot x}{K_m+S}. \tag{12.3}$$

With unlimited nutrient resources, a value μ is constant and equation 12.3 describes the exponential growth of the cell population. With the appearance of any factors limiting this growth, μ will decrease. Most often, the value, which is limiting the growth in microbiological systems, is the concentration of the substrate.

In turn, the monomeric compounds formed in the first stage, under the influence of acidogenic bacteria (second stage), turn into volatile fatty and other organic acids, alcohols, aldehydes, ammonia, hydrogen sulfide, carbon (IV) oxide, hydrogen, and water. The formation of pyruvate in this phase, which subsequently leads to the synthesis of acetic acid (12.4) and formate (12.5), is of particular importance. Their source is numerous sugars to undergo fermentation.

$$C_6H_{12}O_6 \rightarrow 2H_2 + 2C_3H_4O_3 \quad \text{(pyruvate)} \tag{12.4}$$

$$4C_6H_5COOH{+}24H_2O \rightarrow 12CH_3COOH + 8H_2 + 4HCOOH \quad \text{(formate)} \tag{12.5}$$

The organic acids (except for acetic and formic acids), formed under the action of a special group of bacteria-acetogens, are converted into acetic and formic acids, hydrogen, etc. As a result of the first three stages (hydrolytic, acidogenic, and acetogenic), acetate and formate, methanol, methylamine, hydrogen, carbon mono- and carbon dioxide, ammonia, hydrogen sulfide, and phosphorus (V) oxide accumulate in the medium. The process of organic matter biodegradation, due to its fermentation, usually begins under the aerobic conditions (oxygen is used to oxidize the substrate biomass during its hydrolytic decomposition).

Thus, the biomethanogenesis process is a complex biochemical and biophysical phenomenon in which at least three groups of bacteria are involved. At the initial stage, the first group of bacteria turns the complex organic substrates into butyric, propionic, and lactic acids. Then, the bacteria of the second group convert these organic acids into acetic acid, hydrogen, and carbon (IV) oxide. The bacteria of the third group (methanogenic bacteria) reduce carbon (IV) oxide to methane with the absorption of hydrogen, which otherwise can inhibit the development of acetic acid bacteria.

Traditionally, the biogas is obtained from the complex organic substances (fiber, starch, proteins, fats, nucleic acids). In our work, the source of organic substances is the excess biomass of aquatic organisms. Therefore, multicomponent microbial associations are used for methane formation. Along with the methane-forming bacteria in the cultures of *Methylococcus* sp., *Flavobacterium* sp., and *Methylosinus* sp., such associations include the microorganisms that convert the organic substrates into methanol, formate, etc. In the case when the substrate is wastewater, in addition to the methanogenic associations, other microorganisms are involved in the process.

At the final stage of methanogenesis, methyl groups are transferred via the transporters to coenzyme M (KoM-SH), with the formation of methyl-KoM. Then, it is restored, accompanied by the decomposition of the complex and the release of CH4. Both reactions are catalyzed by a methyl reductase system, which is a multienzyme complex. In addition to the enzyme, it contains coenzyme M, factor F430. ATP, Mg^{2+} ions, and also not yet identified cofactors are necessary for the activity of the system. The oxidation of hydrogen by carbon (IV) oxide is catalyzed by the enzymes of the oxidoreductase class, and the decarboxylation of acetate occurs with the participation of lyases only in representatives of two genera – *Methanosarcina* and *Methanothrix*.

The intensity of methane synthesis largely depends on the concentration of oxygen and other inhibitors of this process in the substrate. The ratio of various conditions of the course of methanogenesis (the total reaction equation $4C_6H_5COOH + 18H_2O \rightarrow 15CH_4 + 13CO_2$): pH 6.0–8.0, temperature – at least 30°C, the ratio of N:C and solid components to water, determines its duration from 8 to 20 days (Vavilin, 1986).

The stoichiometry of the process can be described by a simplified reaction equation:

$$(C_6H_{10}O_5)n + nH_2O \rightarrow 3nCH_4 + 3nCO_2. \tag{12.6}$$

It was found that the derivatives of propyl alcohol and propionic acid slow down the release of methane, the reaction is inhibited at the stage of acetic acid formation:

$$R\text{–}CH_2\text{–}CH_2\text{–}OH(COOH) \rightarrow 2CH_3COOH + 4H_2 + CO_2. \tag{12.7}$$

The nitrogen-containing compounds additionally give nitrogen (IV) oxide, although in small quantities:

$$R\text{–}NH_2\text{–}CXHY\text{–}OH(COOH) \rightarrow 2CH_3COOH + 4H_2 + CO_2 + NO_2. \tag{12.8}$$

Actually, methane formation occurs according to the following reactions:

$$(C_6H_{10}O_5)_n + nH_2O(H^+) \rightarrow nC_6H_{12}O_6, \tag{12.9}$$

$$4HCOOH \rightarrow 4H_2 + 4CO_2 \tag{12.10}$$

$$C_6H_{12}O_6 + 2H_2O \rightarrow 2C_3H_3COOH + 4H_2 + 2CO_2 \quad \text{(pyruvate)} \tag{12.11}$$

$$4H_2 + CO_2 \rightarrow CH_4 + 2H_2O \quad \text{(30\% methane)} \tag{12.12}$$

$$CH_3COOH \rightarrow CH_4 + CO_2 \quad \text{(70\% methane)} \tag{12.13}$$

Methanobacteria are characterized by their ability to grow in the presence of hydrogen and carbon dioxide. Under natural conditions, methanobacteria are closely associated with the microorganisms that form hydrogen. Such a symbiosis is beneficial and contributes to the development of hydrogen synthesizers, since methanobacteria are able to consume hydrogen, reducing its amount, methane formation occurs with the symbiotic existence of *Clostridium pectinofermentans* and *Methanosarcina vacuolata*. The substrates of methanogenesis are hydrogen, methanol, and acetate, which are synthesized by *C. pectinofermentans* from pectin. The associations of methane-forming microorganisms are obligate anaerobes, and only a few methanogens can withstand the short-term presence of oxygen. The appropriate conditions in the bioreactor are provided, in addition to sealing the equipment, due to the presence of facultative anaerobic bacteria such as *Escherichia coli*, etc.

To a large extent, the quality of the substrate affects the process of biogas formation. On the one hand, the diverse composition of the methanogenic association of bacteria allows recycling almost any organic waste. On the other hand, to maintain the proper level of vital activity of all microorganisms, it is necessary to regulate the ratio of C:N content in the substrate between 10:1 and 30:1 (according to one source) and 30:1 (according to other sources). If the C:N ratio exceeds 30:1, then a lack of nitrogen slows down the methane fermentation process; if the C:N ratio is less than 10:1, a significant amount of ammonia is formed, which is toxic to bacteria. In practice, the optimal ratio C:N = 30:1 by weight can be achieved by mixing the substrates rich in carbon and nitrogen.

Another relationship is also known for determining fermentation. According to it, fats produce the highest biogas yield with high methane content, proteins produce a slightly lower yield, but also with a high CH_4 content, and carbohydrates produce the lowest yield with a low methane content:

$$a = 0.92 \cdot f + 0.62 \cdot c + 0.34 \cdot b, \tag{12.14}$$

where

f, c, and b – the content of fats, carbohydrates and proteins in 1 g of dry matter, respectively.

The coefficients in the formula mean the specific biogas yield from the starting materials. On the other hand, not only the composition but also the consistency of the substrate affects the performance of the installation. Prevention of its delamination and thorough grinding is important (Ruzhins'ka & Baranova, 2009; Komada, 2019).

Nowadays, 17 genera of methanobacteria are known (*Methanosarcina*, *Methanobrevibacter*, *Methanobacter*, *Methanococcus*, *Methanospirillum*, *Methanothrix*, etc.). Among them, the *Methanospirillum hungati* and *Methanobacter formicicum* species are dominant. Later, Japanese scientists discovered another species of this genus – *M. kadomensis*. Methanogenesis by mass culture of this strain is carried out in 8 days (usually for this process it takes 20 days). The use of a new strain can significantly improve the organization of the biometagenesis process in practice.

All strains of methanobacteria are characterized by the ability to grow in the presence of hydrogen and carbon dioxide. These bacteria also show high sensitivity to oxygen, as well as to other methane formation inhibitors. About 50 species from 17 genera belonging to *Archaebacterio bionta* can synthesize methane. Although it is customary to consider them as a group of methanosynthetic bacteria, these species are genetically rather heterogeneous. Bergey identifies three orders of methanogenic bacteria: *Methanococcales*, *Methanobacteriales*, and *Methanomicrobiales*.

The development of all methanogenic bacteria is sensitive to the oxygen content in the system; for some of them, the growth is completely suppressed even when the oxygen content in the gas phase is at a concentration of 0.004%. Most of the methanogens belong to mesophylls; their optimum development temperature is 30°C–40°C, and the optimum pH is 6.5–7.5 (Sasson, 1984). Almost half of the methanogen species belong to autotrophs; the CO_2 fixation in them is due to acetyl-KoA, whereas the nitrogen fixation is the characteristic of some methanogens (*Methanobacterium formicium*, *Methanosarcina barkeri*).

For the sulfur fixation mechanism, a reduced form is most often needed. It is possible to attract sulfur in the molecular form and in the form of a sulfite ion. Only a small number of species of methanogens (*Methanococcus thermolithrophicum*, *Methanobrevibacter ruminantium*) can absorb sulfur in the form of sulfate ions. The ability to oxidize hydrogen carbon (IV) oxide is characteristic of almost all methanogens. Bryant discovered (Sasson, 1984) the symbiosis of acetic acid and methane-forming microorganisms (previously, it was considered a single microbe and was called *Methanobacillus omelianskii*). A characteristic feature of methanobacteria is the ability to grow in the presence of carbon dioxide and hydrogen. They are also highly sensitive to the oxygen and methane synthesis inhibitors.

Trophic association is beneficial for both methanobacteria and hydrogen-reducing bacteria – in vivo, they are closely interconnected. Methanobacteria use hydrogen gas for live, which is a product of the activity of hydrogen-reducing bacteria. As a result of associative coexistence, the concentration of hydrogen decreases to the threshold of danger for the hydrogen-reducing bacteria.

The studies (Ruzhins'ka & Baranova, 2009; Sasson, 1984) found that the growth of methanogenic bacteria, and accordingly the biogas yield, is suppressed with decreasing pH. A decrease in the biogas emission is also observed with an excessive increase in the degree of loading of the reactor with raw materials. In order to regulate the optimal pH values in the system (the maximum degree of conversion is observed under the conditions close to neutral, the pH is 6.0–8.0), dosage of lime is used. The choice of the optimal temperature range depends on the mode (mesophilic or thermophilic) the biogas synthesis process is realized in. In the first case, it is 30–40°C, in the second – 50°C–60°C, while sharp changes in the temperature regime (the so-called "thermal shock") are not allowed.

The authors of scientific works (Komada, 2019; Malovanyy et al., 2019a,b; Nykyforov, 2016, 2019) investigated the process of methanogenesis of the blue-green algae biomass. Biomethanogenesis never ends with the complete methanization of the biomass of BGA. Less than half of the organic component is involved in bioconversion. If the total potential energy reserve of 1 kg of dry matter of cyanobacteria is 17.75 MJ, then by biomass methanization, it is possible to achieve a biogas yield that is equivalent to 7.54 MJ/kg of dry weight. The residues of unreacted organic matter after methane fermentation can be used as fertilizer in agriculture and forestry (Aires, 1999; Barannikov, 1985; Borisenko, 1982; Pasenko & Nykyfotova, 2018). It was found that methane synthesis takes place in the membranes of bacterial cells and is associated with the generation of a transmembrane potential, the energy of which is transformed into ATP (Figure 12.3).

Considering methanogenesis from a biochemical point of view, it can be imagined as anaerobic respiration, during which electrons are transferred from organic substances to carbon (IV) oxide, which causes the reduction of CO_2 to CH_4 (during normal fermentation, acetaldehyde is reduced to ethanol). The electron donor for methanobacteria is hydrogen, which is produced by several types of hydrogen-reducing anaerobic bacteria. The energy output for each mol of methane formed by methanogens does not exceed two moles of ATP. Therefore, for its growth, methane-forming bacteria must synthesize a significant amount of methane. Out of the 100% metabolized carbon compounds, they transform only 5%–10% into cellular material, the rest is converted to methane. The real effectiveness of methanogenesis can be deduced from the mathematical equation:

$$E = \frac{VCH_{4r}}{VCH_{4\text{th max}}} \cdot 100\% \qquad (12.15)$$

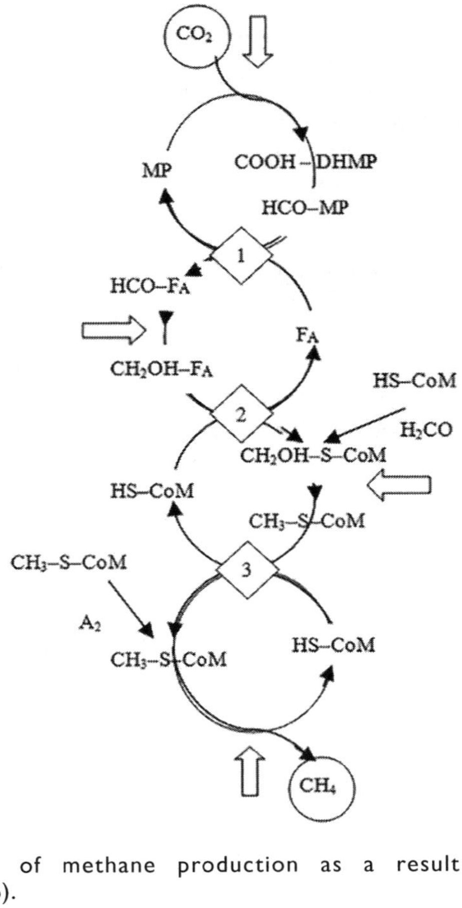

Figure 12.3 The scheme of methane production as a result of reduction CO_2 (Vavilin, 1986).

where

VCH_{4r} – the volume of methane that is actually formed in the digester with the sample of biomass of sludge and wastewater under study, cm^3;

$VCH_{4th\ max}$ – theoretically calculated maximum possible volume of methane, cm^3 (according to Clapeyron's equation):

$$VCH_4 = \frac{NCH_4 \cdot P \cdot T_0 \cdot V_{g \cdot ph}}{T_1 \cdot P_0} \tag{12.16}$$

where

NCH_4 – the methane content in the gas phase, fractions of a unit;

$V_{g \cdot ph}$ – the volume of the gas phase in the digester, dm^3;

T_0 – temperature at n at (298 K);

T_1 – the temperature at which methanogenesis occurs, K;

P_0 and P – the pressure under normal conditions and in the digester, respectively, Pa (Pasenko & Nykyfotova, 2018; Vavilin & Vasiliev, 1979; Vitkovskaya, 2005).

An active mesophilic accumulative culture that stably hydrolyzes cellulose to methane in 5–7 days at a cultivation temperature of 30°C can be relatively easily isolated from activated sludge from the digesters of the biological wastewater treatment plant. From the accumulating methanogenic culture, strains of primary anaerobes are subsequently isolated, which hydrolyze cellulose (cellulolytic bacteria), fermentative (saccharolytic bacteria), and secondary anaerobes – methane-forming archaea (Borisov, 2005; Krig et al., 1997; Shlegel, 1987).

12.2.2 Characterization of the Activated Sludge Biomass Formed in Wastewater Treatment Processes

The main factors in the formation of the biocenosis of sludge from wastewater treatment plants are the composition of the treated effluents and the load on activated sludge for organic substances. The largest group of microorganisms in activated sludge are bacteria. Their number ranges from 10^8 to 10^{12} cells in 1 g of absolutely dry sludge. The composition of the bacterial population of sludge largely depends on the concentration of pollutants in wastewater. About 50%–80% of activated sludge bacteria belong to the *Pseudomonadaceae*. Together with heterotrophic bacteria, active debris always contains active destructors of organic impurities – lower fungi, the mass of which is up to 30% of the sludge biomass. The biocenosis of activated sludge also contains protozoa (important indicator organisms of the quality of sludge), rotifers, worms, and other representatives of the mesofauna. The technical improvement of the systematic hydrobiological control of activated sludge can improve the performance of treatment facilities (Gnida, 2017).

In addition to the biochemical oxidation of wastewater pollution, the microbiocenosis of bacteria and sludge fungi plays an important role in the formation of the physicochemical structure of activated sludge flakes. The ability to form flakes is an important property of activated sludge. It ensures the existence of a spatial structure for the efficient adsorption of wastewater pollution, forms an immobilization surface and an ecological environment for the functioning of the mesofauna of silt. In addition, the ability to form flakes facilitates the separation of the liquid phase (purified wastewater) from the sludge biomass by sedimentation. In mixed cultures, flakes form intensively. The structure of sludge flakes changes with the mass development of filamentous bacteria and some fungi in the activated sludge, which leads to "swelling of activated sludge".

The formation of flakes is associated with the process of endogenous respiration at the metabolic stage, when the ratio of nutrient to prokaryotic biomass in wastewater becomes low. First, the cells oxidize the accumulated substances, then cellular lipids, carbohydrates, and proteins. In turn, the energy level of the activated sludge system also decreases (insufficient supply of movement energy). The energy of motion counteracts gravity, so if it is low, then the reaction is also low. Under these conditions, the prokaryotic cells are mutually attracted to each other. An important factor in their flocculation is the electric charge on the cell surface, which is mainly negative in the pH range of 4–9. Flocculation, the formation of a developed surface and a significant adsorption capacity of flakes are also facilitated by the formation of capsules by bacteria and the release of mucous liquid on the surface of cells.

The chemical composition of mucus and capsules is formed by acetyl groups and amino groups. These groups are responsible for the charge on the surface of the cells: a negative charge is formed when the former dominates and a positive charge when the latter dominates. Indeed, well-known theoretical claims regarding flocculation of activated sludge apply to any cell surface containing amino acids and their polymers, including peptidoglycans, proteins, and glycoproteins (Kvaternyuk et al., 2017; Petruk et al., 2015).

However, if the wall of the bacterial cell and its derivatives (capsule or mucus) are formed only by exopolysaccharides (bacterial starch, hemicelluloses or mucopolysaccharides), then it is more correct to talk about adhesion. The latter is not associated with the formation of partial electric charges in different parts of molecules containing the $-NH_2-$ and $-OOC-$ radicals (Cenens et al., 2000; Martins et al., 2004). Under laboratory conditions, an analysis of the composition of nutrients in wastewater sludge from a sludge site was carried out (Table 12.1).

The data obtained indicate that the content of useful elements in the sediments is high. Its value is close to humus and precipitation may well be disposed of as high-quality fertilizer for agricultural purposes. The samples were taken from the silt site at a depth of 0–0.2 and 0.2–0.4 m. As a result of chemical analysis, compounds

Table 12.1 The content of nutrients in wastewater sludge from a sludge site

Chemical substance	*Component content*
Total phosphorus, %	1.21–4.81
Total nitrogen, %	0.73–0.79
Total potassium total, %	1.43–2.41
Humidity, %	54–84.5
pH	6.6–7.1

Table 12.2 The content of nutrients in wastewater sludge from a sludge site

Parameter name	*Depth of selection, m*	Ni^{2+}	Cu^{2+}	Cr^{6+}	Zn^{2+}	Fe^{3+}
The content of heavy metals, in WWS from the sludge site, g/kg	0–0.2	0.094	0.3	0.055	0.78	30.05
	0.2–0.4	0	0.72	0.075	0.94	80.47
MPC in WWS, g/kg		0.2	0.75	0.5	1.75	0.5

such as ferum (III) oxide were found: nickel (II) oxide, copper (II) oxide, chromium (VI) oxide, and zinc oxide. The data were obtained showing the absence of excess MPC for heavy metals in wastewater sludge (Table 12.2).

12.2.3 The Use of Activated Sludge Biomass in the Wastewater Treatment Processes of Dairy Enterprises

In the production of cheese products, a mandatory accompanying component is milk whey, the quantity of which depends on the type of the product being manufactured and the specific conditions of the individual production. The main physical and chemical indicators of some types of whey are presented in Table 12.3.

The chemical composition of whey varies widely. It relates to the consumption of fat in a normalized mixture. For example, it depends on the type of hard rennet cheese, its fat content, casein whey – on the type of casein (lactic acid, hydrochloric acid, rennet) and the common fat in skim milk. In addition, the serum content depends on the quality of the feedstock, as well as on the type of equipment that is used in a particular technological scheme (Petruk et al., 2012; Wójcik & Smolarz, 2017; Wójcik et al., 2019).

In Ukraine, until recently, whey was practically not processed and was not completely developed together with industrial effluents in the city wastewater system, which is the most dangerous environmental pollution. This is mainly due to highly polluted effluents, which are explained by the features of the technologies used. The characteristics of the wastewater of dairy enterprises are presented in Table 12.4.

The pH value of the wastewater is largely determined by the production technology, the range of products. For the industries not related to the lactic acid fermentation processes, the pH of the wastewater is close to neutral (6.8–7.4 for a dairy cannery, a

Table 12.3 The main physical and chemical parameters of milk whey

Indicators	*Milk whey (milk curd)*			
	From fermented milk cheeses	*From rennet cheeses*	*Casein*	*Co-cited*
Solids content, %	4.2–7.4	4.5–7.2	4.5–7.2	5.3–5.5
Including				
Milk sugar	3.2–5.1	3.9–4.9	3.5–5.2	4.5–4.7
Mineral substances	0.5–0.8	0.3–0.8	0.3–0.9	0.5–0.9
Milk fat	0.05–0.4	0.2–0.5	0.02–0.1	0.02–0.1

Table 12.4 Characteristics of wastewater from dairy enterprises

Indicators	*Value range*
Temperature, °C	25–70
Chemical oxygen demand (COD), mgO/dm^3	2,000–15,000
Biological oxygen consumption (BOD_5), mgO_2/dm^3	800–5,500
Suspended matter, mg/dm^3	300–1,500
Fats, mg/dm^3	100–450
Total nitrogen, mg/dm^3	50–60
Total phosphorus, mg/dm^3	15–2
pH	3.5–11.5

butter-making factory). At the cheese-making factories, city dairy plants and other enterprises producing cheese and dairy products, a certain amount of milk whey is discharged into the wastewater, which leads to a decrease in the pH of wastewater to 6.2.

The total nitrogen content in the wastewater of urban dairies, dairy plants, and creameries is 50–60 mg/dm^3, or 4.2%–6% of total BOD; cheese factories – 90 mg/dm^3, or 3.7% of BOD, the phosphorus concentration is 0.6–0.7% of BOD full. Today, almost all developed countries in the world use methanogenesis as the main stage of purification for the disposal of concentrated effluents from food industry enterprises.

Under laboratory conditions, the wastewater that had a chemical oxygen demand (COD) of 4,200–4,300 mg O_2/dm^3 was cleaned. COD was determined with the titrimetric method. Periodic fermentation of sludge was carried out in a laboratory reactor with a volume of 2 dm^3 under reflux during thermostating. Before the experiment, in a mixture of wastewater and sludge, the pH of the medium, the initial value of COD, and total seeding were determined. In order to increase the efficiency and speed up wastewater treatment before the start of the digestion process, the specially adapted anaerobic activated sludge taken from wastewater treatment plants was introduced into each sample. The amount of sludge introduced in each study was 10% of the volume of wastewater in the reactor.

With a periodic mode of fermentation, the daily loading dose was 30% and 50% of the total volume of culture fluid. Fermentation time for effluent under anaerobic conditions was three days. The fermentation temperature ranged from 25°C to 65°C. The ratio between C:N by weight was provided at 28:1. When carrying out the process at 50°C, which corresponds to the value of the thermophilic temperature range, the following results were obtained. Within 24–25 hours, the formation of biogas is practically not observed, while the cultivation medium was acidified from pH 6.4 to 4.5. Next, the formation of biogas began, which was collected in the corresponding graduated cylinders with the "water shutter" method.

The amount of biogas increased exponentially, and its release ended after 67–72 hours. The specific amount of biogas formed in this case was 3.8 dm^3/dm^3 of wastewater. The analysis of the gas composition was carried out on a "Crystal-2000M" gas chromatograph. The chromatographic analysis of the gas showed that the gas composition changed depending on the phase of development of microorganisms. In the first stages at the initial moment of gas formation, the determination of the biogas composition showed 64% methane, 34% carbon dioxide, as well as the impurities of hydrogen and carbon monoxide CO. In the middle of the phase and at the end of the

Table 12.5 The main indicators of purification and gas generation during periodic fermentation of wastewater of the dairy, depending on the dose

Loading dose	*Final chem. oxygen demand (CODfin)*	*The amount of biogas/ wastewater*	*Content CH_4*	*The degree of wastewater treatment*
%	mgO_2/dm^3	dm^3/dm^3	%	%
30	190	3.8	73	95.3
50	240	3.5	67	94.0

process, the methane content in biogas increased to 70%–73%, and carbon dioxide decreased to 26%–29%, respectively. A change in the pH of the cultivation medium and a slight evolution of gases indicate a two-stage decomposition of organic pollutants. At the first stage, the decomposition of macromolecular compounds into acids – acetic, propionic, butyric – occurs. At the second stage, methane fermentation proceeds with the release of methane and carbon dioxide. It should be noted that the change in COD during the process and gas generation is directly proportional. With the initial COD values of wastewater 4,000 mg O_2/dm^3 at the end of the COD purification process, it was 190–240 mg O_2/dm^3. The results of the research on wastewater treatment in periodic mode are shown in Table 12.5.

An analysis of the data showed that an increase in the loading dose from 30% to 50% with periodic fermentation of effluents led to a decrease in the specific amount of biogas by 8.6% (from 3.8 to 3.5 dm^3/dm^3). With an increase in the loading dose from 30% to 50%, the methane content in biogas in the second experiment decreased by 8.9%. Thus, when periodically fermenting wastewater from dairy plants, it is recommended that before the start of the process, the specially adapted anaerobic activated sludge is added to the digester tank in an amount of 10% and the dose should be 30%.

12.3 CONCLUSIONS

It was established that the main technological problem of wastewater treatment at biological treatment plants is the increased load on activated sludge due to the discharge of effluents with a high concentration of organic pollutants. This leads to the violations of the regulatory process of biological wastewater treatment. As a result, the sedimentation properties of excess activated sludge deteriorate, which affects the slowdown of the compaction and dehydration of wastewater sludge. This necessitates an increase in the treatment time of wastewater, an increase in air consumption for aeration, or the introduction of highly contaminated wastewater from dairies at the pre-treatment facilities for wastewater treatment into the anaerobic treatment cycle.

It was revealed that the excess activated sludge that accumulates at treatment plants does not contain heavy metal compounds in the concentrations that exceed the maximum permissible concentrations for all chemicals.

It was shown that an increase in the loading dose from 30% to 50% with periodic fermentation of dairy wastewater led to a decrease in the specific amount of biogas by 8.6% (from 3.8 to 3.5 dm^3/dm^3). With an increase in the loading dose from 30% to 50%, the methane content in biogas decreased by 8.9%. Thus, when periodically fermenting wastewater from dairy plants, it is recommended that before the start of the process, the specially adapted anaerobic activated sludge is added to the digester in the amount of 10% of the wastewater volume, and the loading dose should be taken equal to 30%.

REFERENCES

Aires, R.M. 1999. *Analiz stochnykh vod dlya sel'skokhozyastvennogo ispol'zovaniya*. Moskva: Meditsina Zheneva VOZ.

Barannikov, V.A. 1985. *Okhrana okruzhayushchei sredy v zone promyshlennogo zhivotnovodstva. Meditsina*. Moskva: Rosselkhozizdat.

Borisenko, E.G. 1982. Biotransformatsiya zhidkikh navoznykh stokov promyshlennykh svinootkormochnykh kompleksov. In Biokonversiya: tez.dokl. Vsesoyuz. simp. Riga: Riga Polytechnical Institute, 15–180.

Borisov, L.B. 2005. Meditsinskaya mikrobiologiya, virusologiya, immunologiya; *Meditsinskoe informatsionnoe agentstvo*. Moscow: 1–734.

Cenens, C., Smets, I.Y. & van Impe, J.F. 2000. Modelling the competition between floc-forming and filamentous bacteria in activated sludge wastewater treatment systems. A prototype mathematical model based on kinetic selection and filamentous backbone theory; *Water Research*, 34(9): 2535–2541.

Epoyan, S.M., Gorban, N.S. & Fomin, S.S. 2010. Analiz sushchestvuyushchikh metodov ochistki stochnykh vod molokozavodov; *Naukovyi visnik budivnitstva*. Kharkiv; *KhDTUBA, KhOTV ABU*, 57: 393–398.

Gnida, A. 2017. Use of DAIME for characterisation of activated sludge flocs; *Environmental Protection*, 43(4): 66–103.

Kezlya, K.O., Tkachenko, T.L., Semenova, O.I. & Bubliienko, N.O. 2011. *Ochishchennya stichnikh vod molokopererobnykh pidpriyemstv perspektivnyi napryam prikladnoi ekologii; Zbirnik tez dopovidei Vseukrainskoi naukovo–praktychnoi konferentsii "Voda v kharchovii promislovosti"*. Odesa: ONAKhT, 144–145.

Komada, P. 2019. *Analiza procesu termicznej przeróbki biomasy*. Warszawa: Monografie – Polska Akademia Nauk. Komitet Inżynierii Środowiska.

Krig, N., Snit, P., Steili, D.S. & Khoult D.G. (red) 1997. *UiliamsOpredelitel' bakterii Berdzhi: v2*. Moskva: Mir: 19–97.

Kuris, Y.V. 2016. Metanogenez i tekhnologichni skhemy otrymannya biogazu; *Energosberezhenie, energetika, energoaudit*, 10(92): 41–48.

Kvaternyuk, S., Pohrebennyk, V. & Petruk, R. 2017. Increasing the accuracy of multispectral television measurements of phytoplankton parameters in aqueous media; *17th International Multidisciplinary Scientific GeoConference SGEM 2017: SGEM2017 Vienna GREEN Conference Proceedings*, Vol. 17(33): 219–225, Vienna, Austria.

Malovanyy, M., Moroz, O., Hnatush, S., Maslovska, O., Zhuk, V., Petrushka, I., Nykyforov, V. & Sereda, A. 2019a. Perspective technologies of the treatment of the wastewaters with high content of organic pollutants and ammoniacal nitrogen; *Journal of Ecological Engineering*. 20(2): 8–15.

Malovanyy, M., Zhuk, V., Nykyforov, V., Bordun, I., Balandiukh, I. & Leskiv, G. 2019b. Experimental investigation of Microcystis aeruginosa cyanobacteria thickening to obtain a biomass for the energy production; *Journal of Water and Land Development*, 43(1): 113–119.

Martins, A.M.P., Pagilla, K., Heynen J.J. & van Loosdrecht, M.C.M. 2004. Filamentous bulking sludge – a critical review; *Water Research*, 38(4): 793–817.

Nykyforov, V., Malovanyy, M., Aftanaziv, I., Shevchuk, L. & Strutynska, L. 2019. Developing a technology for treating blue-green algae biomass using vibro-resonance cavitators; *Naukovyi Visnyk Natsionalnoho Hirnychoho Universytetu*, 6: 181–188.

Nykyforov, V., Malovanyy, M., Kozlovs'ka, T., Novokhatko, O. & Digtiar, S. 2016. The biotechnological ways of blue-green algae complex processing; *Eastern-European Journal of Enterprise Technologies*, 5/10(83): 11–18.

Pasenko, A. & Nykyfotova, A. 2018. Biotechnology of preparation and the use of sediment of wastewater as a fertilizer-meliorant. In Sobczuk, H. & B. Kowaslka (eds.), Water Supply and Wastewater Disposal. Lublin: Monografie, 231–241.

Pashkov, A.P. 2011. Problemi zabrudnennya poverkhnevikh, pidzemnikh i stichnikh vod ta zakhodi shchodo iikh likvidatsii i zapobigannya v Ukraïni; *Bezpeka zhittediyal'nosti*, 4: 10–16.

Pastorek, Z., Kara, Y., Linnik, N.K., Golub, G.A. & Targonya, V.S. 2006. Proizvodstvo biogaza iz smesi organicheskikh materialov; *Naukovii visnik NUBIP Ukrainy*, 95(1): 144–149.

Petruk, V., Kvaternyuk, S. & Denysiuk, Y. 2012. The spectral polarimetric control of phytoplankton in photobioreactor of the wastewater treatment; *Proceedings of the SPIE*, 8698: 86980H.

Petruk, V., Kvaternyuk, S., Yasynska, V., Kotyra, A. & Romaniuk, R. 2015. The method of multispectral image processing of phytoplankton for environmental control of water pollution; *Proceedings of the SPIE*, 9816: 98161N.

Ruzhins'ka, L.I. & Baranova, I.G. 2009. Model protsesu anaerobnogo ochishchennya stichnoi vody v bioreaktory z lystovymy nerukhomymy nosiyami immobilizovanoi mikroflory; *Naukovi visti NTUU "KPI"*, 2: 84–88.

Sasson, A. 1984. *Biotechnologies: Challenges and Promises*. Paris: United Nations Educational, Scientific and Cultural Organization.

Shinkarenko, V.F., Zagirnyak, M.V. & Shvedchikova, I.A. 2010. *Structural-Systematic Approach in Magnetic Separators Design. In Computational Methods for the Innovative Design of Electrical Devices*. Berlin: Springer: 201–217.

Shlegel, G.G. 1987. *Obshchaya mikrobiologiya* (*Allgemeine Mikrobiologie*). Moskva: Mir.

Vavilin, V.A. 1986. *Vremya oborota biomassy i destruktsiya organicheskogo veshchestva v sistemakh biologicheskoi ochistki*. Moskva: Nauka.

Vavilin, V.A. & Vasiliev, V.V. 1979. *Matematicheskoe modelirovanie protsessov biologicheskoi ochistki stochnykh vod aktivnym ilom*. Moskva: Nauka.

Vitkovskaya, S.E. 2005. Izmenenie soderzhaniya podvizhnykh form khimicheskikh elementov v tekhnologi transformatsii organicheskogo veshchestvya komposta; *Agrokhimiya*, 4: 27–31.

Wójcik, W., Pavlov, S. & Kalimoldayev, M. 2019. *Information Technology in Medical Diagnostics II*. London: Taylor & Francis Group, CRC Press.

Wójcik, W. & Smolarz, A. 2017. *Information Technology in Medical Diagnostics*. London: CRC Press.

Zagirnyak, M.V. 1996. *Research, Calculation and Improvement of Pulley Magnetic Separators*. Kiev: ISMO.

Zagirnyak, M.V., Branspiz, Y.A. & Shvedchikova, I.A. 2011. *Magnitnye Separatory. Problemy Proektirovanija*. Kiev: Tehnika.

Chapter 13

Ecological and Economic Principles of Rational Use of Biomass

Oksana A. Ushakova
Technical College of National University of Water and Environmental Engineering

Nataliia B. Savina and Nataliia E. Kovshun
National University of Water and Environmental Engineering

Larysa E. Nykyforova
National University of Life and Environmental Sciences of Ukraine

Natalia V. Lyakhovchenko
Vinnytsia National Technical University

Waldemar Wójcik
Lublin University of Technology

Gulzada Yerkeldessova and Ayaulym Oralbekova
Kazakh University Ways of Communications

CONTENTS

13.1 INTRODUCTION

One of the main issues of the Ukraine's economy is the irrational use of energy sources. The period of industrial development of the world economy was characterized by the large-scale use of natural fossil energy sources in the real sector of social production, which led to their gradual depletion. The fuel and hydrocarbon model of world energy development has lost its relevance; so, in our opinion, further improvement of the national energy policy is possible only if the energy intensity of gross domestic product (GDP) is reduced, economic efficiency of energy distribution and use is optimized, as well as energy saving, development of competitive relations in the markets of energy sources, optimization of production of own energy sources, diversification of external sources of their providing and use of biomass for energy purposes are ensured.

DOI: 10.1201/9781003177593-13

The goal of the European Union's (EU) transition to climate-neutral development by 2050, set out in the European Green Deal strategy, announced by the new European Commission will significantly accelerate the energy transformations in the EU countries, and it will be reflected in all areas of the economy in cooperation with other countries in Europe and the world. In particular, the agreement approved the long-term energy goals aimed at reducing the greenhouse gas emissions, increasing the share of renewable energy sources (RES), in particular from biomass, improving energy efficiency and others (European Green Agreement, Wantuch & Janowski 2014).

According to the Energy Strategy of Ukraine for the period up to 2035, the share of agrobiomass (straw, stems) in 2020 should be 5.3% (Energy Strategy of Ukraine). According to the predictions of the Bioenergy Association of Ukraine (BAU-UABIO) in the structure of biofuel use in Ukraine by 2050, agrobiomass (straw, stems) will be 7.9% (Strategy for the development of bioenergy in Ukraine).

In order to achieve these goals, Ukraine needs a new progressive energy policy focused on ensuring competitiveness, security, and sustainable energy development. This highlights the need to improve the approaches to the formation of ecological and economic principles pertaining to the rational use of biomass in the context of the need to achieve in Ukraine the level of 5.3% in the structure of gross final energy use in 2035, which determined the research.

The purpose of the study is theoretical and methodological justification and development of applied recommendations for the formation of the ecological and economic principles of rational use of biomass. Achieving this goal led to the solution of the following tasks: conducting a systematic analysis of the scientific and methodological foundations of the state energy policy in the context of greening the economy; improving the system of ecological and economic forecasting of the state of the environment (NPS) based on the consequences of using agricultural lands for growing energy crops in the region; improving the methodological approaches to the rational distribution of biomass on the basis of substantiation of ecological and economic expediency of using the straw of grain crops for energy needs taking into account the quality of soils; improving the functional structure of the logistics system for managing the material flows of collection, use and redistribution of the straw of cereals as energy raw materials in the region; substantiating the amount of thermal energy that can be replaced by heat from the straw of cereals as RES, taking into account the balance of humus in the soil; and determining the annual ecological and economic effect in Rivne region as a result of energy substitution.

13.2 MATERIAL AND RESEARCH RESULTS

The authors found that the sustainable development of society, the manufacturing sector and, in particular, the energy sector should be harmonized with the environmental component; to ensure the unity of nature management and the realization of energy, economic, and environmental interests of the state. Energy efficiency is realized on the basis of the principles of innovation and greening of production where the main strategic directions of rational development are 'green' economy and innovative experience of the European economic development strategy 'Europe 2020: a strategy for smart, sustainable and inclusive growth'.

Given the Ukraine's participation in the Energy Community, the priorities of energy policy are to focus on ensuring competitiveness, security, and sustainable energy development. Accordingly, we propose the components of implementing the state energy policy in the context of greening the economy, which should be based on the implementation of world experience and the principles of 'green' economy (Figure 13.1). Thus, based on the provisions of the 'green' economy, it was determined that the scientific and methodological principles of state energy policy should be based on environmental and economic tools for sustainable energy supply and land use to increase the energy potential of RES, including biomass.

Ukraine is one of the most energy-intensive economies in the world. According to the Statistical Yearbook of World Energy 2019 in 2018, the energy intensity of domestic GDP was 0.238 kep/$ in 2015 prices (Statistical Yearbook of World Energy 2019).

Standard GDP does not take into account the quality of NPS (Official site of the State Agency for Energy Efficiency and Energy Saving of Ukraine). GDP grows from the increase in pollution and from the cost of eliminating the damage caused. The real value of GDP in the context of the sustainable development paradigm and the 'green' economy should be adjusted to the environmental factor of economic efficiency of natural resources. The 'green' indicator (environmentally adjusted) GDP is not included

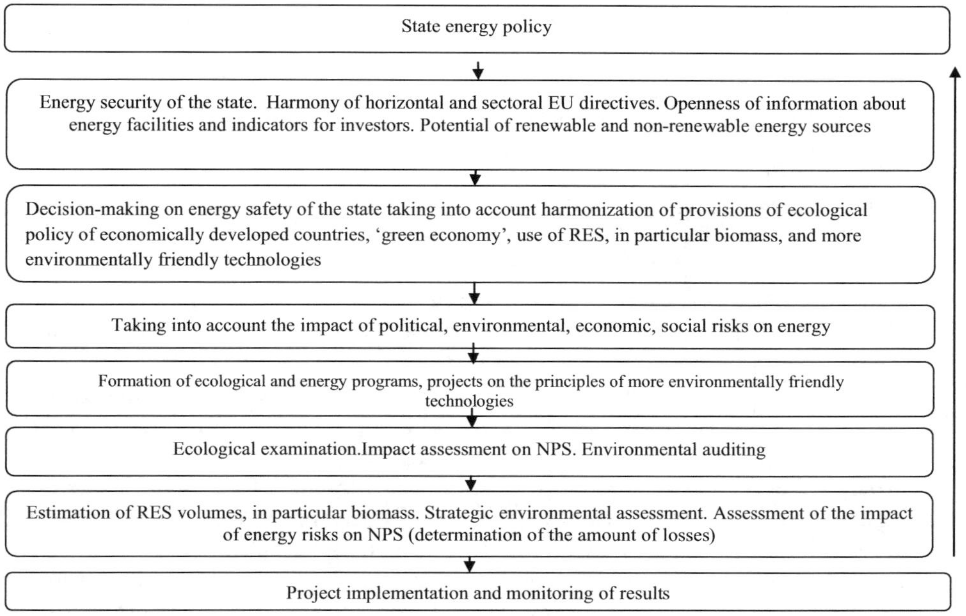

Figure 13.1 The components of implementing the state energy policy in the context of greening the economy.

in the system of national accounts. Therefore, this indicator should be determined by the formula (13.1) (Pashechko 2014):

$$\text{'Green' GDP} = \text{GDP} - \text{SPR} - \text{DNPS} - \text{VONPS} - \text{CE}, \tag{13.1}$$

where

SPR – consumption of natural resources (reduction of natural resources);
DNPS – degradation of NPS (damage due to economic activity);
VONPS – costs for the protection of NPS;
CE – unresolved synergistic effects from balanced (sustainable) nature management (authors' proposal).

The authors understand the obtained synergetic effects as the positive socio-ecological and economic effects of the realization of the potential of biomass available for energy use, which will result in 'green' GDP. In addition, our country is one of the countries with a shortage of its own fossil fuels.

An urgent problem in the world is the contradiction between ensuring the food security through the cultivation of agricultural products and the use of agricultural land for the cultivation of energy crops. In the context of Ukraine's high level of energy dependence, the authors substantiate the feasibility and importance of using the vegetable waste from agriculture (biomass) to replace the fossil energy sources to strengthen the energy security of the state.

According to the State Agency for Energy Efficiency and Energy Saving, it has been established that the annual energy technically achievable potential for the production of energy from renewable sources and alternative fuels in Ukraine is over 98.0 million tons/year, in particular bioenergy – 31.0 million tons/year, through the implementation of which Ukraine can save natural gas and reduce energy dependence of the state (Official site of the State Agency for Energy Efficiency and Energy Saving of Ukraine). According to the Energy Balance of Ukraine, the share of biofuels in the total supply of primary energy (ZPPE) in 2018 amounted to 3.2 million tons. BC is 3.4% of ZPPE (Strategy for the development of bioenergy in Ukraine, Nikiforova 2015).

According to BAU-UABIO, the biomass in Ukraine is enough to replace all imports of gas, coal, and gasoline as of 2018. The energy potential from biomass is 23 million tons. The structure of biomass available in Ukraine for energy purposes in 2018 is as follows: grain straw and rapeseed – 4.04 million tons. n., wood biomass – 3.08 million tons. n., by-products of corn for grain and sunflower – 5.1 million tons. n., sunflower husk – 1.0 million tons. n. (Strategy for the development of bioenergy in Ukraine, Shtovba 2007).

The processes of commercialization of domestic bioenergy are underdeveloped, so the fuel and energy complex, of course, lacks a number of economic benefits. In Ukraine, it is advisable to use the experience of European countries in the development of bioenergy through increasing the land fertility and crop yields.

In view of this, it is necessary to develop domestic agriculture for the production of energy crops and their use for commercial purposes, taking into account the quality of soils. It is important to determine the impact of renewable biomass cultivation on NPS. The authors propose a system of ecological and economic forecasting of the state of NPS on the consequences of the use of agricultural land for growing energy crops in the region (Figure 13.2).

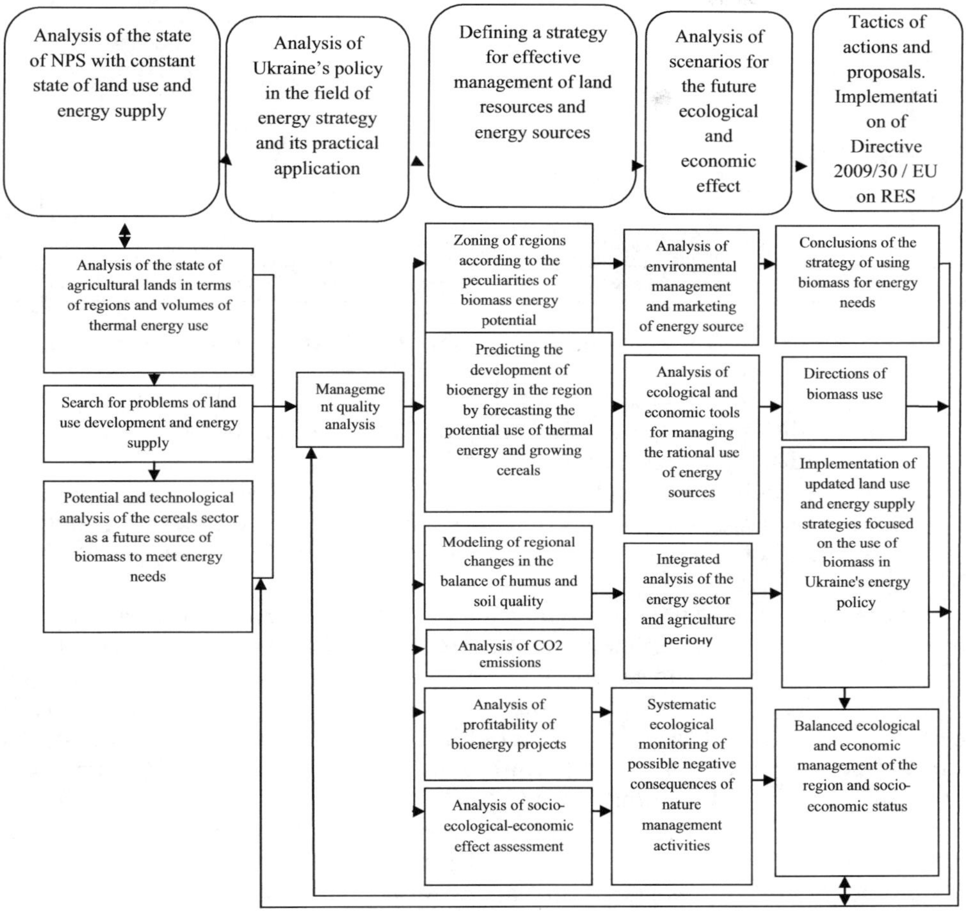

Figure 13.2 The system of ecological and economic forecasting of the state of the environment based on the consequences of the use of agricultural lands for growing energy crops in the region.

The system allows:

- to analyze the state of NPS with the state of energy supply and land use, which will positively affect the quality of management of energy sources and land resources;
- to determine the scenarios of obtaining socio-ecological-economic effect on the basis of zoning of regions according to the peculiarities of biomass energy potential;
- to forecast the volume of heat use and cultivation of grain crops;
- to calculate regional changes in the balance of humus and soil quality;
- to improve the tools of environmental management, logistics, and marketing of energy sources.

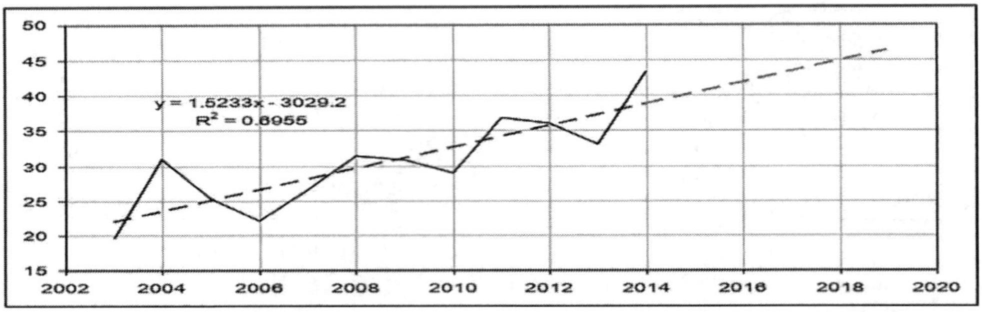

Figure 13.3 Forecast of winter wheat yield for Rivne region for the period up to 2020, c/ha.

Under the conditions of the existing energy potential, system of agriculture, development of productive forces, natural-climatic and geographical conditions the use of biomass is promising in the Rivne region (Ukraine), where there is a tendency to stable growth of winter wheat yield. According to the forecast results obtained by the authors, the level of yield of this grain crop in the region in 2020 may reach 47 c/ha (Figure 13.3).

At the same time, while maintaining the traditional option of plowing straw into the soil, a negative balance of humus at the level of −1.86 c/ha should be expected. Taking into account the ecological and agrochemical requirements for maintaining soil fertility according to the options for applying winter wheat straw to agricultural fields, the authors proved that the improvement of humus balance results in up to 0.18 c/ha with additional application of 12.5% straw to the soil; up to 2.22 c/ha – 25%; up to 6.03 c/ha – 50%; and up to 10.38 c/ha – 75% (Pashchenko 2014).

As for the humus content, after 5 years, according to the fifth option, its content will be 1.667%, and after 10 years – 1.734% (Pashchenko 2014) (Figure 13.4).

On the basis of the results of calculations of the humus balance according to the options of plowing winter wheat straw into the soil and forecasting the yield of winter

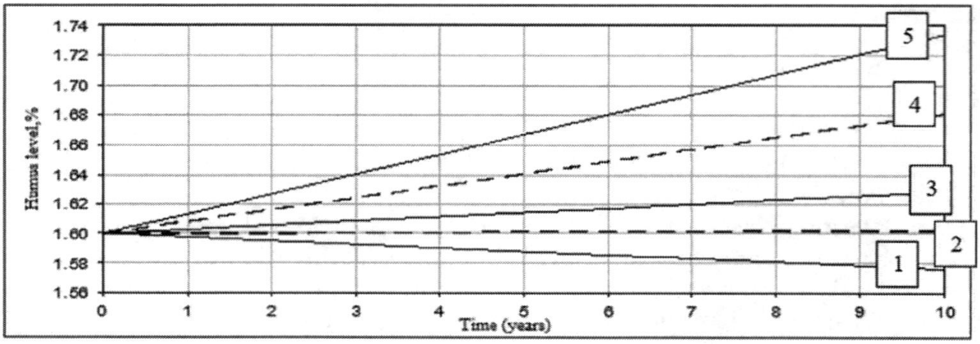

Figure 13.4 Dynamics of humus content in the soil depending on the amount of plowed straw, %.

wheat, the authors determined that the mechanism of using straw crop residues gives a double effect: ecological – improving the soil quality by increasing the share of humus and economic crops as a result of increasing the humus content. The remaining volumes of straw should be used for energy purposes as renewable biomass.

The traditional approach in the EU is to use 20% of straw for the energy purposes.

On the basis of the results of substantiation of expediency of growing energy raw materials, there is a need to develop a logistics system for managing material flows of collection, use and redistribution of straw cereals as energy raw materials in the region (Figure 13.5).

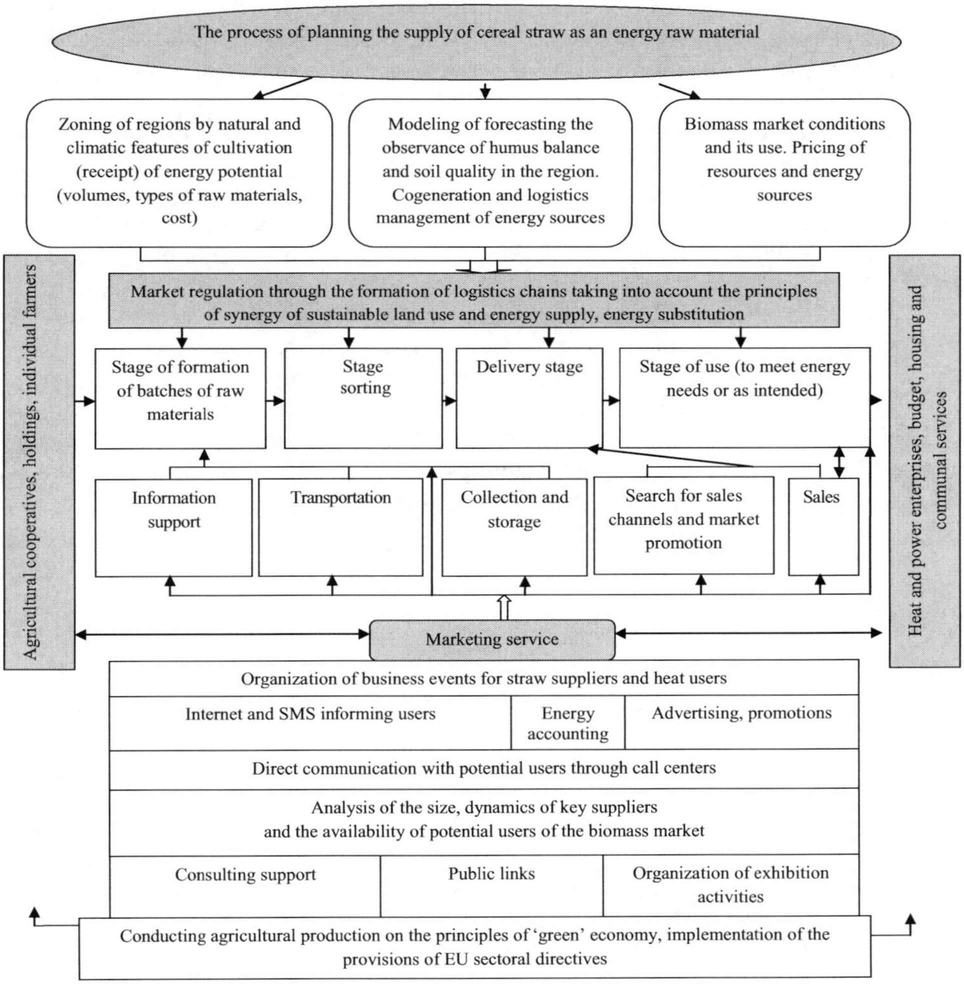

Figure 13.5 Features of the functional structure of the logistics system for managing material flows of collection, use, and redistribution of straw of cereals as energy raw materials in the region.

Under the notion logistic control system by material streams of straw of grain crops, we mean the set of interrelated organizational elements of the market, such as planning logistics, management of material flows of straw of grain crops and the organization of marketing service, within our proposed functional structure of intra- and interregional logistic systems.

As a result of using the proposed structure based on the principles of synergy of sustainable energy and land use, energy substitution is possible to maintain the agricultural production on the principles of green economy, the implementation of horizontal and sectoral EU directives. This will improve the system of protection of the NPC, to promote the integration of the national and regional economies into the global market, the development of socially responsible ('humanitarian') energy.

Therefore, according to the system of environmental-economic forecasting of the NPC based on the results of the agricultural land use for growing energy crops in the region (Figure 13.2), the functional structure of the logistics system for materials management collect, use and redistribution of straw of grain crops, as an energy resource in the region (Figure 13.5), it is advisable to apply comprehensively to improve the ecological and economic mechanism of rational use of renewable energy in the region (Pashchenko 2015a,b, Lytvynenko et al. 2019).

Given the methodological approaches to the rational allocation of biomass, the authors justified the amount of thermal energy from traditional fossil energy sources, which can be replaced with the heat obtained from the combustion of biomass, in particular the straw of grain crops in the Rivne region (Table 13.1).

The authors estimate that if a positive balance of humus is achieved, the value of the annual ecological and economic effect of the use of biomass as RES can range from 6,194.7 to 121,881.7 thousand UAH, depending on options. The most environmental-friendly and in line with the European approaches, as well as economically acceptable, is the option of using 25% of cereal straw for energy needs. It does not provide the largest amount of annual savings, but provides a sufficient level and the highest level of humus balance in the soil. The greatest ecological and economic effect is obtained by using 50% of straw for the energy needs.

13.3 CONCLUSIONS

Thus, the proposed scientific and methodological approach to determining the environmental and economic efficiency of biomass uses a balance of energy needs and fertility indicators of agricultural land, taking into account the effect of replacing traditional fossil energy sources with renewable biomass. As a result, its application creates positive economic, environmental, and social consequences for the development of bioenergy on the principles of "green" economy, which will contribute to the greening of energy under the economic conditions of European integration, will increase the agricultural production, some of which will be used for energy needs, sustainable land use, and conservation of the NPS resources.

Table 13.1 Ecological and economic effect from the use of biomass as a renewable energy source in the Rivne region in 2018

Indices	*Options for the amount of straw that can be used as a renewable energy source*			
	87.5%	*75%*	*50%*	*25%*
Estimated volume of thermal energy, Gcal (according to the State Statistics Service of Ukraine)	1,775,512	1,775,512	1,775,512	1,775,512
The cost of thermal energy without the use of biomass, thousand UAH (at the cost of Ukrainian coal 3,966 UAH/t)	1,005,954.4	1,005,954.4	1,005,954.4	1,005,954.4
Volume of thermal energy to be produced from a traditional energy source, Gcal (depending on the share of replacement of traditional thermal energy with energy from biomass combustion) (Pashechko 2014)	130,849.4	307,308.7	659,689.3	1,012,973.6
The cost of thermal energy produced from a traditional energy source, thousand UAH (at the cost of Ukrainian coal in Ukraine 3,966 UAH/t)	74,135.5	174,112.3	373,761.1	573,921.9
Volume of thermal energy to be produced from biomass, Gcal (depending on the share of replacement of traditional thermal energy by energy from biomass combustion) (Pashechko 2014)	1,644,662.6	1,468,203.3	1,115,822.7	762,538.4
The cost of the amount of thermal energy to be produced from biomass, thousand UAH (at the average market price of straw of cereals in Ukraine 1665 UAH/t)	622,082.4	555,337.9	422,052.3	288,425.0
The cost of thermal energy produced taking into account the effect of energy substitution, thousand UAH	696,218.0	729,450.2	795,813.5	862,346.9
The level of the economic effect (the amount of annual savings), thousand UAH	309,736.4	276,504.1	210,140.9	143,607.5
Ecological effect (humus balance), c/ha (depending on the forecast of humus balance according to the options of plowing straw into the soil, Figure 13.4) (Pashechko 2014)	0.18	2.22	6.03	10.38
Coefficient of environmental friendliness of calculation options	0.02	0.21	0.58	1
Ecological and economic effect, thousand UAH	6,194.7	58,065.9	121,881.7	143,607.5

REFERENCES

European Green Agreement. https://uk.wikipedia.org/wiki/

Lytvynenko, V. et al. 2019. Development, validation and testing of the Bayesian network of educational institutions financing. *2019 10th IEEE International Conference on Intelligent Data Acquisition and Advanced Computing Systems: Technology and Applications (IDAACS)*, 412–417. doi:10.1109/IDAACS.2019.8924307

Nikiforova, L.O. 2015. A generalized model for assessing the level of motivated threat agents in the tasks of ensuring the security of objects at the micro and macro levels. *Suchasnyi zakhyst informatsii (Modern Information Protection)* 4: 71–76.

On approval of the Energy Strategy of Ukraine for the period up to 2035 'Safety, energy efficiency, competitiveness': Order of the Cabinet of Ministers of Ukraine No. 605-r of 18.08.2017/ The Verkhovna Rada of Ukraine. Official site of the Verkhovna Rada of Ukraine. https://zakon.rada.gov.ua/laws/show/605-2017-%D1%80

Pashchenko, O.A. 2014. Ecological and economic problems of energy raw materials use. Problems of rational use of socio-economic and natural resource potential of the region: financial policy and investments. *Collection of Scientific Works* XX(4) Kyiv, SEU/Rivne, NUVGP, 187–193.

Pashchenko, O.A. 2015a. Ecological and economic mechanism of energy resources management of the region. *Actual Problems of Economy* 11(173): 240–246.

Pashchenko, O.A. 2015b. Institutional support for optimizing the use of energy resources from biomass. *Actual Issues of Economy* 4: 268–275.

Potential. Official site of the State Agency for Energy Efficiency and Energy Saving of Ukraine. https://saee.gov.ua/uk/activity/vidnovlyuvana-enerhetyka/potentsial

Shtovba, S.D. 2007. *Design of Fuzzy Systems by Means of MATLAB*. Moscow: Hotline-Telecom.

Statistical Yearbook of World Energy. 2019. Intensity of energy use per unit of GDP at constant purchasing power parity (PPP). https://yearbook.enerdata.ru/total-energy/world-energy-intensity-gdp-data.html

Strategy for the development of bioenergy in Ukraine. Bioenergy potential. https://uabio.org/bioenergy-transition-in-ukraine/

Wantuch, A., & Janowski, M. 2014. Whether RES are competitive for conventional sources?. *Informatyka, Automatyka, Pomiary w Gospodarce i Ochronie Środowiska*, 4(4), 105–108. doi:10.5604/20830157.1130208

Chapter 14

Fallen Leaves and Other Seasonal Biomass as Raw Material for Producing Biogas and Fertilizers

Mykhailo O. Yelizarov, Alona V. Pasenko and Volodymyr V. Zhurav
Kremenchuk Mykhailo Ostrohradskyi National University

Leonid K. Polishchuk
Vinnytsia National Technical University

Andrzej Smolarz
Lublin University of Technology

Yedilkhan Amirgaliyev and Orken Mamyrbaev
Institute of Information and Computational Technologies CS MES RK

CONTENTS

14.1 INTRODUCTION

Green plantations of cities and other settlements, gardens, and agricultural lands are a powerful source of plant waste. Every year in autumn, there is a local accumulation of biomass, which reaches too large volumes for their natural biodegradation in the places of formation (Hrubnyk et al. 2017). Most often in Ukraine, they are disposed of in the cheapest way, namely, incinerated, composted or taken to a landfill. Burning the fallen leaves and dry grass pollutes the atmosphere, poses a danger to the human health and creates the risk of fires (Sonko et al. 2017, Hrubnyk et al. 2019).

During the combustion of one ton of plant waste, about 9 kg of smoke microparticles are released into the air, consisting of dust, nitrogen oxides, carbon monoxide, heavy metals, and a number of carcinogenic compounds. In addition, benzopyrene, which has the ability to cause cancer in humans, can be released from the fallen leaves that do not have the access to oxygen. It is known that dioxins are released into the air with smoke as one of the most toxic substances for humans (Razanov & Tkachuk 2015, Hrubnyk et al. 2019). Burning leaves on the territory of residential buildings, squares and parks is prohibited by the current legislation (Derzhavni sanitarni normy

DOI: 10.1201/9781003177593-14

ta pravyla utrymannia terytorii naselenykh mists). However, this does not solve the problem, as incineration elsewhere remains a source of emissions of the same toxic components.

Decomposition of the fallen leaves biomass in a natural way depends on the humidity of the environment and is more than 2 years, and its disposal in landfills requires significant costs and increases their area (Dyakonov et al. 2016). Composting of vegetable waste is a rather long process and takes place over several months. In addition, there is a problem of allocating large areas of settlements for the location of compost heaps.

The vegetable waste from urban areas should now be considered as an energy raw material, which will solve the environmental, social and economic problems (Popyk 2014, Resuieva 2015, Dyakonov et al. 2016). Renewable energy sources from biomass are an important component in the energy balance of the world (Hengeveld et al. 2016, Grando et al. 2017). In the countries of the European Union, about 14% of the total energy is produced annually from biomass (Hrytsai & Masliukova 2019).

Domestic and foreign researchers have paid considerable attention to the problems and prospects of biogas production as an alternative energy source in their scientific achievements (Sereda & Cherniavskyi 2013, Henning et al. 2014, Binkovska & Shanina 2015). It is advisable to use vegetable waste and by-products of the agro-industrial sector as raw materials for biogas production (Popyk 2014, Tkachenko et al. 2018, Zapalovska & Bashutska 2019).

As a result of anaerobic fermentation of organic matter by methanogenic associations of microorganisms, fermented biomass is formed, which can be used as a valuable organic fertilizer with a high content of nutrients to improve the soil fertility, also in cities (forest parks, parks, squares) (Stepanenko & Proskurnia 2012, Binkovska & Shanina 2015, Kozak & Okhota 2018).

The use of balanced biofertilizers after methane fermentation improves the physical and mechanical properties of the soil and increases the crop yields by 30%–50% (Makarenko et al. 2014). It was investigated that the use of biofertilizers not only has a positive effect on the productivity of agricultural plants but also does not have a negative impact on the agroecosystem (Drukovanyi & Dyshkant 2013). Ukraine has significant reserves of raw materials that provide the opportunity to produce high-quality biological fertilizers for organic crop production (Belas et al. 2003, Bubliienko et al. 2016).

Despite the huge potential of plant waste, insufficient attention is paid to their disposal. In this regard, the study on the utilization of plant waste, in particular fallen leaves, by anaerobic fermentation to obtain liquid products is a topical issue and needs further investigation.

The current level of development of civilization requires significant amounts of energy consumption, including oil and gas, the reserves of which are being intensively reduced. Therefore, the search for alternative energy sources is an important and urgent task. In this regard, it is advisable to create the installations capable of converting renewable (scattered in nature) energy – water, wind, the Sun – into the energy localized in the right place and in such forms that would be convenient for its use. For example, solar energy is converted into electrical energy in semiconductor solar cells (Bogoboyashchii & Izhnin 2000, Bogoboyashchii et al. 2001, 2005, 2006), in generators powered by Stirling motors, and wind energy in wind generators (GOST 5542–87).

Natural and effective storage and converters of solar energy are plants. In the process of photosynthesis, plants turn inorganic substances (primarily CO_2 and H_2O) scattered in the atmosphere into organic substances that can generate heat during combustion. However, another way is possible to release the energy stored by plants, that is, this is biogas. Technically, it is more profitable and convenient to burn biogas than firewood, dead wood, etc.

According to the previously performed calculations of specialists, more than 120 million tons of organic waste by dry weight are generated annually in Ukraine. Processing such a quantity of waste can produce only biogas from 36 to 75 billion m^3, or in terms of methane, from 20 to 45 billion m^3/year. At present, Ukraine produces less than 20 billion m^3 of natural gas with a demand of about 70 billion m^3. Using at least part of the waste potential will reduce the gas purchases, solve a number of environmental problems, and obtain high-quality fertilizers in addition.

The purpose of the work was to study the possibilities of utilizing seasonal biomass, such as plant materials from urban parks, the result of which would be the production of biogas and organic fertilizers.

14.2 MATERIAL AND RESEARCH RESULTS

Biogas is formed during the decomposition of organic substances as a result of an anaerobic microbiological process – methane fermentation. Depending on the type of organic raw material, the composition of biogas may vary, but the basis is methane (CH_4) – the metabolic product of the methane bacteria and carbon dioxide (CO_2) – a respiration product of microorganisms (Malofeev 1998). In general, biogas includes methane, carbon dioxide, a small amount of hydrogen sulfide, ammonia, hydrogen, and other gases. The composition of biogas determines its physical properties (e.g. volumetric heat of combustion), and, accordingly, the possibility of its practical use.

A variety of organic waste is used as raw materials for the industrial production of biogas. Biogas plants can be installed as treatment facilities at farms, poultry farms, sugar and spirit plants, meat processing plants, and in wastewater treatment in megacities (Elinov 1989, Marinenko 2003, Blagutina 2007). A biogas plant can replace a veterinary sanitary plant, that is, carrion can be disposed of in biogas instead of producing meat and bone meal. Technologies make it possible to obtain biogas from virtually any raw material of organic origin. For example, in Western Europe, more than half of all poultry farms are biogas-heated. Volvo and Scania produce biogas-fueled buses.

For the formation of biogas from plant materials, it is necessary, first of all, to create comfortable anaerobic conditions for the life of three types of bacteria. These bacteria are successors among themselves, that is, the subsequent species eats the waste products of the previous species. Firstly, these are hydrolytic bacteria (they are responsible for the destruction of biomass under the dissociative effect of water and temperature), the second type is acid-forming bacteria (they make it possible to obtain molecules of organic acids from hydrolyzed products) and, finally, methane-forming ones that regulate the processes of consumption of organic acids and biogas formation (Pasenko 2012). In the process of biogas production, not only the bacteria of the methanogen class are involved but all three species.

Ecologically important (and not satisfactorily resolved in Ukraine) is the problem of disposal of leaves fallen from trees in urban parks, weeds, etc. Natural processes of decomposition of leaf biomass are slowed down (Figure 14.1a and b) and, depending on the humidity of the environment, take more than 2 years (Elinov 1989). Utilization of the plant biomass in waste collectors requires significant costs, and the burning of such raw materials leads to air pollution and is prohibited by law. Therefore, in our opinion, the most appropriate solution to the problem of plant biomass utilization (fallen leaves, weeds, etc.) would be to obtain biogas.

In connection with the foregoing, it was relevant to determine the conditions necessary for such a process of decomposition of fallen leaves, the products of which would be combustible gas and organic fertilizer. It should be noted that animal raw materials for biogas production, for example, cattle manure, are a fairly homogeneous and sufficiently crushed mixture. The methane-producing bacteria are initially found in the gastrointestinal tract of cattle. Significant advantages of plant materials are its availability (e.g. fallen leaves in urban parks, weeds) and the cost-effectiveness of biogas production.

In September–October, fallen leaves were collected as plant materials in the Prydniprovskyi park of the city of Kremenchuk. The plant mixture in equal proportions contained leaves of maple, poplar, ash, aspen, and about 10% ragweed. The relevance of recycling ragweed and other allergenic weeds is increasing every year, since the number of people suffering from allergies grows as well. Therefore, one of the tasks set by us in the course of research is the possibility of not just disposing of allergenic plant materials.

In order to destroy the integrity of the wax coating on the leaves and increase the area of interaction for bacteria, the raw material was crushed to the size of small leaves (3–5 cm^2). The total mass of raw materials was about 5 kg. Fallen leaves are generated in an oxygen atmosphere, as a rule, washed with rains; therefore, unlike algae, they do not contain anaerobic bacteria on their surface. In order to colonize the nutrient medium with anaerobic bacteria and optimize moisture, silty (non-flowing) water from

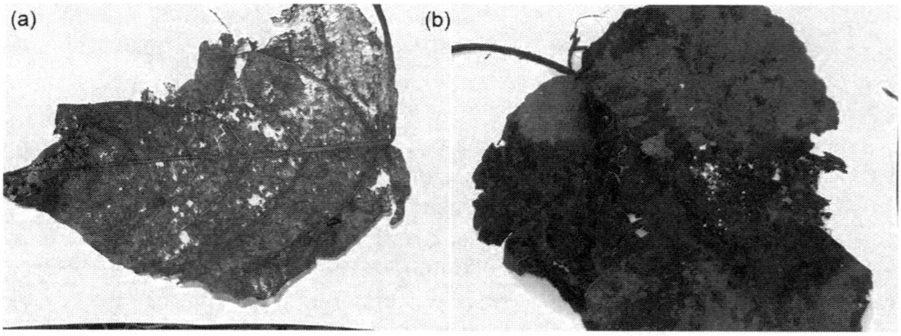

Figure 14.1 Decomposition of leaves under natural conditions for 6 months in a moist (a) and dry environment (b).

the Kremenchug reservoir was added to the plant mass in the volume of 3.5–4 dm^3. For this purpose, wastewater can also be used (Elinov 1989, Pasenko 2019, Pasenko & Nykyforova 2019). No catalysts were added.

The methane bacteria show their vital activity in the temperature range from 3°C–4°C to 70°C–90°C. If the temperature is higher, they begin to die, at sub-zero temperatures they survive, but cease their activity. Since the metabolic activity and the level of reproduction of methane bacteria is lower than that of acid-forming ones, with an excess of acids (pH below 6.5), the activity of methane bacteria decreases.

On the other hand, an increase in pH over 8.0 (excess amino acids) dampens the methane-formation process. Typically, the pH is kept constant. Moreover, for the active life of organisms, it is necessary to maintain normal pressure, for example, to accumulate biogas in an elastic reservoir. The moisture content of the feedstock should be 85%–92% (Blagutina 2007, Nykyforov et al. 2019).

There are two main modes of decay – mesophilic (at a temperature of 25°C–40°C) and thermophilic (at a temperature above 40°C) (Blagutina 2007). For our research, the mesophilic mode was chosen as the most common one. The moist plant mixture was placed in a glass container with a volume of 10 l and placed in the darkness under a water shutter, blocking the access of air. For 9–10 weeks, the temperature of the plant mixture was maintained around the clock in the range of 28°C–32°C. The pH of the resulting mixture was 6.5–7.0. The biogas released during digestion process is accumulated in a special rubber chamber. About 1.5–10 dm^3 of biogas under normal atmospheric pressure was released from the indicated mass of the plant mixture (about 5 kg) within 1.5 months.

The specific heat of combustion of the obtained biogas was determined experimentally using a reference heat sink. The indicated value is determined mainly by the methane content, since insignificant amounts of hydrogen and hydrogen sulfide practically do not affect this indicator. For the biogas obtained, the specific heat of combustion was 9.6 MJ/kg, or 2.3 kcal/l. For comparison: coke oven gas – 16 MJ/kg, natural gas – 35 MJ/kg (Elinov 1989, Malofeev 1998).

The heat of combustion of biogas containing 70% methane is 25.1 kJ/l, or 5.99 kcal/l (Elinov 1989). It is possible that the optimization of the regime of decay will increase the percentage of methane, and, therefore, the heat of combustion of biogas.

It should be noted that the biogas energy use in comparison with the combustion of natural gas, liquefied gas, oil, and coal is CO_2-neutral, since the emitted carbon dioxide is within the natural carbon cycle and is consumed by plants during the growing season. Thus, the concentration of carbon dioxide in the atmosphere, compared with the use of solid fuels, does not increase. In addition, the energy obtained from biogas belongs to renewable energy, since it is produced from an organic renewable substrate (Petruk et al. 2012).

The main areas of biogas use are: (1) power generation; (2) biogas burning in boiler plants for heating water and supplying it to consumers; (3) preparation of biogas in accordance with the requirements of regulatory and technical documentation and its supply to gas distribution networks of local consumers of natural gas (mixing with natural gas); (4) cleaning, compression, and refueling with gas of gas-filled cars, tractors, and other types of transport; and (5) obtaining nutrient biomass as fertilizer.

The choice of biogas use is determined in each case for each farm. The most urgent issue today is providing farms with electricity and burning biogas in water heaters.

In the process of obtaining biogas from plant materials, in particular, from fallen leaves, a digested mass is formed, which can and should be used as a high-calorie organic fertilizer. The previously performed scientific studies (Petruk et al. 2015) show that the resulting biomass does not contain weed seeds, helminth eggs, and other pests. Depending on the composition of the feedstock, on the type of soil, as well as on the crop under which fertilizing is planned, the resulting plant substrate can be enriched with nitrogen, phosphorus, and potash. At the same time, the fertilizer consumption will be 1–5 tons/ha, instead of 50–60 tons of raw manure (Elinov 1989). The quality of the resulting fertilizers ultimately increases the productivity of the soil and reduces the cost of acquiring expensive mineral fertilizers. In some farms, it is more profitable to produce and sell the resulting fertilizer than meat or milk. The environmental quality of agricultural products grown using such fertilizers should also be noted. The chemical composition of digested plant biomass was investigated using a high-precision atomic absorption spectrometer with a continuous spectrum radiation source of the Contr AAR 700 type. The results of the analysis are shown in Table 14.1.

An important feature of the obtained substrate is the absence of heavy metals in the composition, perhaps this is a consequence of the collection of raw materials not in the industrial zone of the city. For a more detailed analysis of the correspondence of the content of heavy metals in raw materials depending on the district of the city and general conclusions, further studies are required.

The resulting substrate was used as a nutrient mixture for germination of seeds of vegetable crops. In several containers with soil mixtures, seeds of common vegetable crops were sown, 20 per each container. For comparison, the same seeds were germinated in ordinary soil (soil taken in urban areas) without a nutrient mixture and other additives.

Ceteris paribus, seed germination in the presence of a nutrient mixture is 100% (Figure 14.2a), in ordinary soil – 40%–70% (Figure 14.2b), the growth and maturation periods also differ. For example, seeds of cucumbers, watermelon, and pumpkin sprouted 1–2 days earlier when plant substrate was introduced into the soil. When transplanting seedlings into open ground, the first sprouts (Figure 14.2a) were more resistant to hot, dry weather.

Thus, biogas can be successfully obtained from the plant materials of park tree species, weeds, in particular, from fallen leaves, as well as from other seasonal biomass. In fact, it is an environmental-friendly fuel similar in its characteristics to natural gas. The calorific value of the resulting biogas indicates the efficiency of its production. The remaining substrate can be used as an effective fertilizer. The use of such raw materials

Table 14.1 The content of certain metals in the plant substrate obtained

Chemical element	*Concentration, mg/l*	*Chemical element*	*Concentration, mg/l*	*Chemical element*	*Concentration, mg/l*
Aluminum	0.07	Cobalt	0	Sodium	1,006.25
Barium	2.48	Silicon	81.25	Nickel	0
Vanadium	0.05	Magnesium	393.75	Lead	0
Iron	1.09	Copper	0	Chromium	0
Potassium	1,968.75	Molybdenum	0	Zinc	0

Figure 14.2 Pumpkin springs, sprouted with the addition of substrate to the soil (a) and in ordinary soil (b).

enables to simultaneously solve energy, and environmental, and social issues, both local and more general in scope.

Firstly, the production of biogas from fallen leaves promotes the efficient utilization of such raw materials and improves the environmental situation. The utilization of leaves with the method of mesophilic digestion followed by the burning of biogas reduces the emission of greenhouse gases.

Secondly, with rising energy prices, the implementation of the project for the extraction of biogas from fallen leaves will become an alternative to traditional energy facilities and will begin to be profitable in a few years.

Thirdly, the development of biogas energy solves the employment problems and contributes to the development of energy infrastructure.

14.3 CONCLUSIONS

The use of fallen leaves as raw materials can simultaneously address the energy, environmental, social, and economic issues, as the implementation of a project to extract biogas from fallen leaves can be an alternative to traditional energy, and the development of biogas energy solves employment problems and promotes energy infrastructure.

Of course, it is too early to talk about such large-scale transformations in the energy sector in Ukraine, and such forecasts seem unlikely. This is primarily due to the lack of a well-established legislative system for assessing and regulating relations in this area, as well as government mechanisms to stimulate the introduction and use of alternative energy sources. The creation of new enterprises for the processing of exclusively fallen leaves is impractical due to the seasonal nature of the receipt of this type of resource. Therefore, fallen leaves should be considered as an additional resource on existing plants, rather than as the main raw material, or use it in combination with other seasonal biomass (agricultural waste, etc.).

Another positive consequence of the disposal of fallen leaves is the reduction of the required area in landfills for solid waste. The total area of landfills and dumps in Ukraine is about 6,000 ha, and their number exceeds 3.5 thousand, while the majority of them are in unsatisfactory sanitary and epidemiological condition, and constitute the objects of increased environmental danger.

The use of fallen leaves as a secondary raw material provides the following positive changes in the management of solid waste:

- minimizing the accumulation of fallen leaves in urban landfills and dumps;
- reduction of the negative impact of fallen leaves on the environment and human health;
- reduction of costs for removal of fallen leaves to landfills and solid waste landfills;
- ensuring the extension of the service life of landfills, by reducing the volume of removal of fallen leaves for further disposal.

Reducing the working area of municipal solid waste landfills through the disposal of fallen leaves is a promising initiative, both from an environmental and economic point of view. This is due to the fact that the maintenance of landfills requires large amounts of funding and the elimination of environmental consequences – even more so.

The analysis of methodological approaches to the treatment of fallen leaves makes it possible to identify such basics as:

- storage and disposal at landfills and landfills for solid waste;
- composting;
- briquetting.

The promising options for the use of fallen leaves as a secondary resource include:

- anaerobic fermentation with subsequent production of biogas and fertilizers;
- processing of fallen leaves to obtain a sorbent of petroleum products;
- the use of fallen leaves as a filler for gypsum boards.

Thus, the above-mentioned methods of handling fallen leaves allow considering it as a secondary plant raw material, expanding the use and application of fallen leaves in order to reduce the integrated eco-destructive impact on the environment and further economic benefits.

REFERENCES

Belas, E., Bogoboyashchii, V.V., Grill, R., Izhnin, I., Vlasov A. & Yudenkov, V. 2003. Time relaxation of point defects in p- and n-(HgCd)Te after ion milling. *Journal of Electronic Materials* 32(7): 698–702.

Binkovska, H.V. & Shanina, T.P. 2015. Agricultural plant residues in the Odessa oblast: perspectives for biogas production. *Ukrainskyi hidrometeorolohichnyi zhurnal* 16: 107–112.

Blagutina V.V. 2007. Bioresources. *Chimiya i zhizn* 1: 36–39.

Bogoboyashchii, V.V. & Izhnin, I. 2000. Mechanism for conversion of the type of conductivity in p-Hgl–xCdxTe crystals upon bombardment by low-energy ions. *Russian Physics Journal* 43: 627–636.

Bogoboyashchii, V.V., Izhnin, I. & Mynbaev, K. 2005. The nature of the compositional dependence of p–n junction depth in ion-milled p-HgCdTe. *Semiconductor Science and Technology* 21(2): 116–123.

Bogoboyashchii, V.V., Izhnin, I., Mynbaev, K., Pociask, M. & Vlasov, A. 2006. Relaxation of electrical properties of n-type layers formed by ion milling in epitaxial HgCdTe doped with V-group acceptors. *Semiconductor Science and Technology* 21(8): 1144–1149.

Bogoboyashchii, V.V., Vlasov A. & Izhnin, I. 2001. Mechanism for conversion of the conductivity type in arsenic-doped p-Cd x Hg1–x Te subject to ionic etching. *Russian Physics Journal* 44(1): 61–70.

Bubliienko, N.O., Semenova, O.I., Zhylyk, A.V., Semenova, O.A. & Tymoshchuk, T.M. 2016. Bioconversion of vegetable waste agriculture using methane fermentation. *Visnyk Zhytomyrskoho natsionalnoho ahroekolohichnoho universytetu* 2(56): 31–37.

Derzhavni sanitarni normy ta pravyla utrymannia terytorii naselenykh mists [State Sanitary Rules and Rules for Maintaining the Territories of Settlements]. 2011. No. 145. Retrieved from https://zakon.rada.gov.ua/laws/show/z0457-11 [in Ukrainian].

Drukovanyi, M.F. & Dyshkant, L.V. 2013. Tehnological lines on payment of biogas and biological organic fertilizer for grown environmentally clean agricultural products. *Naukovi pratsi Instytutu bioenerhetychnykh kultur i tsukrovykh buriakiv* 19: 139–143.

Dyakonov, V.I., Dyakonov, O.V., Skrypnyk, O.S. & Nikitchenko, O.Y. 2016. Economic and ecological issues of utilization fallen foliage in city. *Komunalne hospodarstvo mist* 129: 51–55.

Elinov, N.P. 1989. *Chemical Microbiology*. Moscow: Vysshaya shkola.

GOST 5542–87. Natural gases for industrial and municipal use.

Grando, R.L., Antune, A.M., Fonseca, F.V., Sanchez, A., Batrena, R. & Font, X. 2017. Technology overview of biogas production in anaerobic digestion plants: a European evaluation of research and development. *Renewable and Sustainable Energy Reviews* 80: 44–53.

Hengeveld, E.J., Bekkering, J. & van Gemert, W.J.T. 2016. Biogas infrastructures from farm to regional scale, prospects of biogas transport grids. *Biomass and Bioenergy* 86: 43–52.

Henning, H., Krautkremer, B. & Hartmann, K. 2014. Review of concepts for a demand-driven biogas supply for flexible power generation. *Renewable and Sustainable Energy Reviews* 29: 383–393.

Hrubnyk, A.V., Kostohryz, A.P. & Martynuk, K.S. 2017. Research of efficiency of the use of leafy biomass of green plantations of cities as energy source. *Visnyk KhNTU* 4(63): 39–43.

Hrubnyk, O., Podolskyi, M. & Lilevman, I. 2019. Background of use hardwood biomass and plant waste for energy supply in agriculture. *Tekhnikotekhnolohichni aspekty rozvytku ta vyprobuvannia novoi tekhniky i tekhnolohii dlia silskoho hospodarstva Ukrainy* 24(38): 360–368.

Hrytsai, A.H. & Masliukova, Z.V. 2019. Otsinka enerhetychnoho potentsialu biohazu Ukrainy [An estimation of the energy potential of biogas of Ukraine]. *Scientific Horizons* 10(83): 63–58.

Kozak, K.V. & Okhota, Y.V. 2018. The main trends of efficient use of biogas in Ukraine. *Efektyvna ekonomika* 4. doi:10.32702/2306-6814.2019.21.54.

Makarenko, N.A., Bondar, V.I. & Borshch, H.M. 2014. Ecotoxicological assessment of biofertilizers, fermentation products, biogas plants for their compliance with organic production. *Visnyk Poltavskoi derzhavnoi ahrarnoi akademii* 4: 20–24.

Malofeev, V.M. 1998. *Biotechnology and Natural Protection*. Moscow: Arktos.

Marinenko, E.E. 2003. *Principles of Biofuel Production and Use in Order to Solve Problems of Energy-Saving and Natural Protection in Municipal and Agricultural Industries*. Volgograd: VolgGASA.

Nykyforov, V., Sakun, O., Yelizarov, M., Pasenko, A. & Maznytska O. 2019. Test-object activity and mortality depending on electromagnetic radiation intensity and duration. *Materials of the International Conferences on Modern Electrical and Energy Systems*: 380–383, Kremenchuk, Ukraine.

Pasenko, A.V. 2012. Ecological aspect of waste treatment schemes for thermal power plants. *Ekolohichna bezpeka* 2/2012(14): 29–32.

Pasenko, A.V. 2018. Energy-saving technology for the production of ameliorant fertilizer from sludge waste of heatpower companies. In Energy Efficiency and Energy Saving: Economic,

Technical, Technological and Ecological aspects/Collective Monograph. Poltava: Poltava State Agrarian Academy, 322–327.

Pasenko, A.V. & Nykyforova, E.A. 2019. Biotechnology of preparationy and of the use of sediment of sewage as a fertilizer-meliorant. *Water Supply and Wastewater Disposal/Monograph*: 231–241.

Petruk, V.G., Kvaternyuk, S.M., Denysiuk, Y.M. & Gromaszek, K. 2012. The spectral polarimetric control of phytoplankton in photobioreactor of the wastewater treatment. *Proceedings of the SPIE* 8698: 86980H.

Petruk, V.G., Kvaternyuk, S., Yasynska V., Kozachuk, A., Kotyra, A., Romaniuk, R.S. & Askarova N. 2015. The method of multispectral image processing of phytoplankton for environmental control of water pollution. *Proceedings of the SPIE* 9816: 98161N.

Popyk, O.V. 2014. Ecological and economic aspects fallen leaves treatment in urban. *Ekonomichni innovatsii* 58: 266–272.

Razanov, S. & Tkachuk, O. 2015. The comparative analysis of pollutant emissions is into the air by traditional energy and different types of biofuels. *Silske gospodarstvo ta lisivnitstvo* 1: 152–160.

Resuieva, N.S. 2015. Perspectives of using plant waste for generating bioenergy in Ukraine. *Ekonomika: realiyi chasu* 4(20): 179–185.

Sereda, L. & Cherniavskyi, M. 2013. Collecting biogas and liquid bio-fertilizers during biomass processing at mobile machine. *Naukovi pratsi Instytutu bioenerhetychnykh kultur i tsukrovykh buriakiv* 19: 158–162.

Sonko, S.P., Pushkarova-Bezdil, T.M., Sukhanova, I.P., Vasylenko, O.V., Hurskyi, I.M. & Bezdil, R.V. 2017. The problem of utilization of felling leaves of cities and wastes of animal farming farm and ways of its solutions. *Problemy neoekolohii* 1–2(27): 143–154.

Stepanenko, D.S. & Proskurnia, T.O. 2012. Receiving and utilization of biogas from waste. *Pratsi TDATU* 9: 134–143.

Tkachenko, T.V., Yevdokymenko, V.O., Kamenskykh, D.S., Filonenko, M.M., Vakhrin, V.V. & Kashkovskyi, V.I. 2018. Processing vegetable waste of different origin. *Nauka ta innovatsii* 2: 51–66.

Zapalovska, A. & Bashutska, U. 2019. The use of agricultural waste for the renewable energy production. *Naukovi pratsi Lisivnychoi akademii nauk Ukrainy* 18: 138–144.

Chapter 15

Toxicity by Digestate of Methanogenic Processing of Biomass

Volodymyr V. Nykyforov, Dmitrii M. Salamatin, Sergii V. Digtiar, and Oksana A. Sakun
Kremenchuk Mykhailo Ostrohradskyi National University

Leonid K. Polishchuk
Vinnytsia National Technical University

Róża Dzierżak
Lublin University of Technology

Maksat Kalimoldayev and Yedilkhan Amirgaliyev
Institute of Information and Computational Technologies CS MES RK

CONTENTS

15.1 INTRODUCTION

Recently, a very promising biotechnology for the production of methane-containing biogas mixture based on enzymatic fermentation of organic matter of biological origin has appeared on the energy market. This technology certainly solves a number of problematic issues in energy, agriculture and some other areas of economic activity (Gorova & Kulina 2008.). Given the fact that organic waste is used for the production of the target product (primarily livestock), this method of producing biogas can undoubtedly be considered economical and environmental-friendly. However, in order to consider this biotechnology really waste-free, there is a need to dispose of the substrate spent after the process of methanogenesis, which has already undergone fermentation destruction. This substance – digestate – can be used as an organo-mineral fertilizer in both agriculture and forestry. This is convincingly indicated by the results of chemical analysis of various substrate samples (Zakharenko et al. 2017; Nykyforov et al. 2008, 2011; Zagirnyak et al. 2016).

DOI: 10.1201/9781003177593-15

As a result of miscalculations during the construction and subsequent operation of the cascade of reservoirs on the Dnieper River, there was a problem of man-made chemical pollution of surface natural waters in the region with specific organic substances (Zuman et al. 2013; Malovanyy et al. 2016). In this regard, considerable attention is paid to the research aimed at preventing or limiting organic pollution, in particular due to the massive development of blue-green algae (BGA) in artificial reservoirs. The outbreaks of "blooms" increase the level of toxicity of natural waters, worsen the regime of reservoirs, as well as suppress the life of aquatic organisms and inhabitants of adjacent habitats. The annual seasonal process of "blooming" and the subsequent death of aquatic organisms determine the development of technological options for solving the environmental problem (Nykyforov & Kozlovs'ka 2002). Today, there are about 40 members of the genera of toxigenic cyanobacteria, including *Microcystis*, *Anabaena*, *Nodularia*, *Nostoc*, *Aphanizomenon*, *Oscillatoria*, and *Cylindrospermopsis*. However, the main accumulator of organic matter during the "blooming" of the Dnieper is a representative of photosynthetic cyanobacteria – *Microcystis aeruginosa* Kützing. It accounts for up to 90% of biomass in the "blooming" spots – the largest cluster of cyanide cells in the reservoir.

Therefore, a digestate-based digestate or a combined mixture with their addition, which could potentially contain algotoxins and hazardous cell breakdown products, requires a careful study for the presence or absence of toxic effects on living organisms. In connection with the above, the removal of BGA from the reservoirs of the Dnieper cascade (Lugovoy et al. 2007) and the use of their biomass for further processing will not only obtain an additional source of nutrients (Nykyforov et al. 2016; Digtiar 2016), but also improve the quality of natural surface waters and drinking water in particular, which is supplied to the populated cities of coastal areas. The problem of freeing water bodies from excess BGA can be considered as one that will allow the targeted use of natural biomass producers, which contains valuable food, feed, medical, pharmaceutical, perfume, agricultural, and forestry important components (Nykyforov et al. 2016, Abiru et al. 2015).

15.2 MATERIALS AND METHODS

In order to study the possibility of using spent digestate based on BGA and other substrates after methanogenesis as biofertilizer on the example of test crops – *Triticum aestivum* L. and *Pisum sativum* L., the following groups of methods were used: chemical, instrumental, mathematical methods in biology, and growth test.

Detection of the degree of acute toxic effect of the tested water is based on the national standard "Water quality. Determination of acute lethal toxicity on *Daphnia magna* Straus and *Ceriodaphnia affinis* Lilljeborg (*Cladocera*, *Crustacea*)" ISO 6341:1996, MOD DSTU 4173:2003. To do this, the graphical method was employed to calculate: LC_{50} – 24 hours – multiplicity of dilution of the tested water, which kills 50% of daphnia in 24 hours; LC_0 – 24 hours – the minimum multiplicity of dilution at which daphnia do not die in 24 hours (Natsionalny standart Ukrainy).

The logarithms of the multiplicities of dilutions of the tested water are plotted on the abscissa axis, and the arithmetic mean values of daphnia survival as a percentage of control are plotted on the ordinate axis. The resulting points are connected by a straight

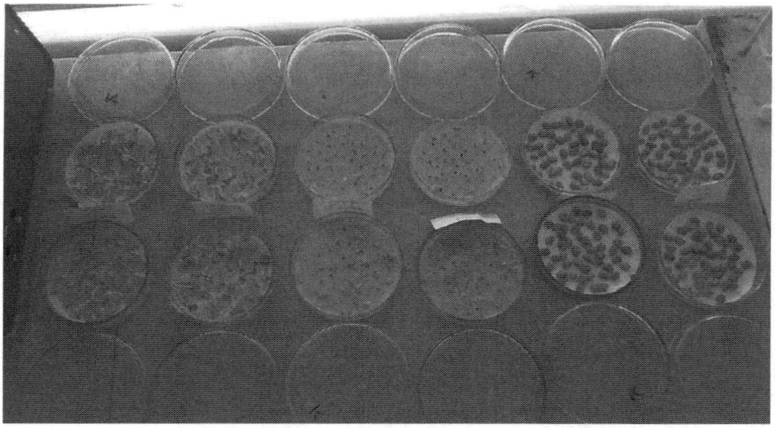

Figure 15.1 Growth test for wheat, mustard and pea seeds.

line. From the points on the y-axis corresponding to 50% and 100% survival, draw lines parallel to the abscissa. From the points of intersection of these lines with the experimental line, the perpendiculars are lowered to the abscissa axis and the logarithms of the dilution multiplicity values are found, which will correspond to the required values of LC_{50} and LC_0. The higher the values of LC_{50} and LC_0, the more toxic the tested water.

The degree of toxicity can also be determined by calculating the LT_{50} – the average time of death of 50% daphnia in the tested water. To do this, build a graph (on the abscissa axis postpone the observation time, on the ordinate axis – survival as a percentage of control). The lower the LT_{50} value, the more toxic the test water. The results of biotesting by this method were summarized in the relevant protocol tables.

The assessment of soil toxicity after the use of digestate as a biofertilizer was carried out using the "Growth test". Out of the several options available: (1) germination of plants on samples of the studied substrates; (2) watering plants in sand or soil culture with the studied liquids; and (3) water culture of plants on extracts from soils, wastes, sewage, etc.; the latter was chosen.

Various plants were used as test crops: wheat, rye, barley, oats, corn, cabbage, radishes, mustard, peas, soybeans, etc. (Figure 15.1). In order to study the soil samples, 100 g of substrate moistened to 70% was applied to each of the experimental tanks, and 15–20 germinated seeds of the test culture were sown. In this case, the indicator can be any plant. The study of all variants was conducted in at least three replicates.

An essential condition of the experiment is to maintain a constant moisture content of the studied soil (at the level of 70% of the total soil moisture content), which is achieved as follows:

- before laying the experiment, the soil is dried and weighed;
- the soil prepared in this way is moistened with such an amount of water that allows reaching 70% humidity;
- moistened in this way the soil is carried into experimental containers and determine the total weight.

Figure 15.2 Samples of germinated seeds in a thermostat.

In order to comply with the above-mentioned conditions, the germination of seeds of the studied plants was carried out in a special thermo-cabinet (Figure 15.2).

When examining the quality of water samples and water extracts, laboratory beakers were filled with test water (250–500 cm^3). For the first few days, the containers with the test samples were covered with glass. Two or three times a day, the glass was removed for 10–15 minutes for ventilation. On the fourth day, the containers with the seeds planted in them were placed on a shelf, where if possible for 14 hours (from 6:00 to 20:00) was maintained constant lighting, and kept under such conditions for another 2 weeks, observing the following indicators:

- – growth dynamics of sprouts (every day);
- – the length of the aboveground part of the shoots and their growth (every day);
- – the total number of germinated seeds (at the end of the experiment).

At the same time, it is necessary to pay attention to morphological features of plants (early yellowing, features of development of root system, etc.).

In 2 weeks, the young plants were carefully released from the soil (or water), shaken off the soil particles and slightly dried on filter paper. Then, the root and stem vegetative systems were measured, the wet weight of sprouts was determined in each of the three replicates, after which the plants were placed in paper bags and dried in an oven at $t^\circ = 70°C–90°C$ to constant weight, and then the dry weight was determined. For each of the investigated options, the average length of the aerial and root systems was calculated using the formula $\bar{x} \pm m$, where m – arithmetic mean error, which is determined by the formula (Azarov et al. 2015; Osadchuk & Osadchuk 2011):

$$m = \sqrt{\frac{\sigma^2}{N}}, \qquad (15.1)$$

where

N – number of results,

σ^2 – dispersion, which is determined by the formula:

$$\sigma^2 = \frac{\sum_{i=1}^{N}(x-\bar{x})^2}{N}, \tag{15.2}$$

The reliability of the arithmetic mean difference (t) is calculated by the Student–Fisher test:

$$t = \frac{\bar{x}_1 - \bar{x}_2}{\sqrt{m_1^2 + m_2^2}}, \tag{15.3}$$

where

$\bar{x}_1$ – the arithmetic mean of the indicator in the control;

$\bar{x}_2$ – the arithmetic mean of the indicator in the variant;

m_1 – arithmetic mean error in control;

m_2 – the same in the variant.

The phytotoxic effect is determined as a percentage of plant weight, length of root or stem system, number of damaged plants or number of seedlings. Given the amount of plant mass formed, the phytotoxic effect is calculated by the formula:

$$\Phi E = \frac{M_0 - M_x}{M_0} \cdot 100 \tag{15.4}$$

where

M_0 – mass of plants in a container with control soil (water);

M_x – mass of plants in a container with the studied soil (water).

The difference between the arithmetic means is considered significant in value $t \geq 3$. The results of arithmetic mean calculations $\bar{x} \pm m$, dispersion σ^2 and the values of the difference of the arithmetic mean t are presented in Table 15.5.

The mathematical methods used during the research include statistical methods (Lakin 1990). Among them, numerous biometric methods of quantitative evaluation of research objects were used, as biometrics is a section of variation statistics, which is used to process experimental data and observations, as well as to plan the quantitative experiments in biological research. During the work, the absolute and percentage of objects (cells, individuals, taxa of different levels) were most often determined.

When biotesting natural water samples, the percentage of dead daphnia in the tested water was calculated and compared with the control, according to the formula (Pavlov et al. 2010; Valtchev et al. 2016; Vasilevskyi et al. 2017):

$$A = (x_c - x_t) \times 100 / x_c, \tag{15.5}$$

where

x_c – the arithmetic mean of the number of daphnia surviving in the control,
x_t – the arithmetic mean number of daphnia surviving in the tested water.

If $A \geq 50\%$, the tested water has an acute toxic effect, and if $A < 50\%$, the tested water has no acute toxic effect on daphnia.

15.3 RESEARCH RESULTS AND DISCUSSION

Before the start of the research, a review of the literature data on the characteristics of the spent cyanobacterial substrate, its biochemical composition and the degree of reproducibility, systematic position and distribution in nature of methane-forming bacteria, their morphological, cytological, and biochemical features and main industrial strains was conducted.

In order to study the properties of the organic matter of cyanide in the upper reaches of the Dniprodzerzhynsk Reservoir within the city of Kremenchuk, water concentrate with a total weight of about 10 kg was selected. The species composition and the number of cells in the selected substrate were determined using a Goryaev camera and a light microscope.

In order to obtain a dry residue, the liquid concentrate was exposed for 96 hours at an average daily temperature of +28°C. Drying was performed in shallow flat-bottomed cuvettes with a volume of 0.5–1.5 dm^3. As a result, the dry mass samples were obtained on average 18 g from 1 dm^3 of natural concentrate, which allows referring the area of the reservoir where the samples were taken to the fifth class of flowering as hyper-flowering (Agadzhanyan et al. 1997). In order to determine the mineral content of a sample weighing 10 g in the obtained dry mass, three portions weighing 10 g were burned in a muffle furnace PM – 8 at a temperature of 800°C for 40 minutes. After complete combustion of organic matter, the weight of the samples was 1.7, 2.0 and 2.0 g, respectively (average 19% of dry weight).

The samples of the spent substrate remaining after the cessation of the methanogenesis process were studied separately. According to X-ray microstructural analysis, it does not contain heavy metals (Nykyforov et al. 2018) and, therefore, can be used as a biofertilizer for feeding any plants, including food crops.

The analysis of the biotesting methods was carried out, namely carrying out of the growth test and features of use of plants as bioindicators. The characteristics of the component composition of biofertilizer and the method of determining the phytotoxic effect of the spent substrate of BGA after methanogenesis are given. The paper uses comprehensive biotesting techniques to detect acute (lethal) toxicity of natural and manmade waters with the use of lower crustaceans as test objects, as well as a growth test according to a unified method, which involves the use of seeds of typical crops.

The biological methods of research were employed to determine the toxicity of substrates using biotesting. Biotesting is a mandatory element of the water quality assessment and control system. The purpose of biotesting is to verify compliance of the water quality with the regulatory requirements. The studies were performed according to the basic method of short-term biotesting, the results of which concluded that there was no acute toxic effect of the studied water samples on the test objects.

The samples of the test liquid were taken taking into account the requirements of ISO 5667-2:1991 "Water quality. Sampling. Part 2. Guidelines for sampling (DSTU ISO 5667-2:1991)". Actually, the biotesting procedure was performed no later than 6 hours after their selection. Before biotesting, the samples were filtered through a filter paper with a pore size of 3.5–10 µm. If acute toxicity was detected, the samples were tested without dilution. For control (water without toxic substances), tap water was used, which was dechlorinated with settling and aeration with a microcompressor for 7 days.

The procedure of hydrotoxicological analysis was carried out in accordance with the mandatory method of short-term biotesting, according to which typical aquatic organisms, inhabitants of local aquatic ecosystems, representatives of the class *Crustaceae* (daphnia or ceriodaphnia) are used as test objects. In particular, in these studies, the test object was a laboratory culture of crustaceans *Daphnia magna* Straus (*Cladocera* order), which is a typical mesosample and can withstand salinity up to 6% and lowering the oxygen concentration in water to 2 mg/dm^3.

For the manifestation of the state of toxic toxicity, test samples were tested without dilution. The volume of the test drives without dilution for biotesting – 500 cm^3. Посуді poured from 10 previous vessels of 15 cm^3/skin (+repeated three times), and the vessels were filled with the same amount of control water (also three times repeated). At the skin from the previous and control vessels, one instance of daphne was perceived. They were transported from the vessels for cultivation by means of a tube with a sawed edge of a diameter of 5–7 mm and exponentially with optimal brains with a length of 24 years. For optimal conditions, it is necessary to ensure that the concentration of acid in water is not less than 7 mg/dm3, the water temperature is 20°C ± 2°C, and the light intensity is 400–600 lux/day of 12–14 year. At the time of the short-hour testing, the year of the Daphne was kicked.

Using standard methods of biotestation, concentration was indicated, and for such test cases, gostro toxic toxicity was recognized, as for a digestate on the basis of fermented biomass less than green greens (Table 15.1), as well as more (Table 15.2).

Table 15.1 The results of digestate biotesting based on cyanobacterial biomass (using *Daphnia magna* Straus as test objects)

Dilution	*Survival, %*			
	Digestate		*Substrate*	
	Absolute number	*%*	*Absolute number*	*%*
Test object – Daphnia magna Straus				
1:10	3	10.0	2	6.6
1:50	24	80.0	1	3.3
1:100	30	100.0	5	16.6
1:200	30	100.0	24	80.0
1:500	30	100.0	27	90.0
1:1,000	30	100.0	27	90.0
Control	30	100.0	30	100.0

Table 15.2 The results of biotesting of digestate based on a mixture of biomass of cyanobacteria and activated sludge (using *Daphnia magna* Straus as test objects)

Dilution	*Survival, %*			
	Digestate		*Substrate*	
	Absolute number	*%*	*Absolute number*	*%*
Test object – Daphnia magna Straus				
1:10	5	16.6	7	23.6
1:50	30	100.0	10	33.3
1:100	30	100.0	18	60.0
1:200	30	100.0	28	93.3
1:500	30	100.0	30	100.0
1:1,000	30	100.0	30	100.0
Control	30	100.0	30	100.0

These results are also confirmed by the data obtained during the test for seed germination. The corresponding calculations are given in the reporting tables (Tables 15.3 and 15.4) and graphs (Figures 15.3 and 15.4).

According to the method of Gorova AI, the spent cyanobacterial substrate (digestate) was air-dried and diluted in boiled water in 0.03, 0.05, 0.07 and 0.09 g/10 cm^3 of boiled water. In Petri dishes on filter paper, 20 grains of *Triticum aestivum* L. were placed and filled with solutions with a certain concentration of digestate and water (control sample).

Table 15.3 Wheat germination (%) at various dilutions of the spent substrate (biofertilizer)

Dilutions	*1*	*2*	*3*	*Average value*
1	2	3	4	5
1:10	89	87	88	88.0
1:50	88	91	87	88.7
1:100	89	96	92	92.3
1:200	97	98	95	96.7
Control	98	99	98	98.3

Table 15.4 Pea germination (%) at various dilutions of the spent substrate (biofertilizer)

Dilutions	*1*	*2*	*3*	*Average value*
1	2	3	4	5
1:10	78	86	79	81.0
1:50	83	91	90	88.0
1:100	94	93	95	94.0
1:200	88	86	87	87.0
Control	97	98	99	98.0

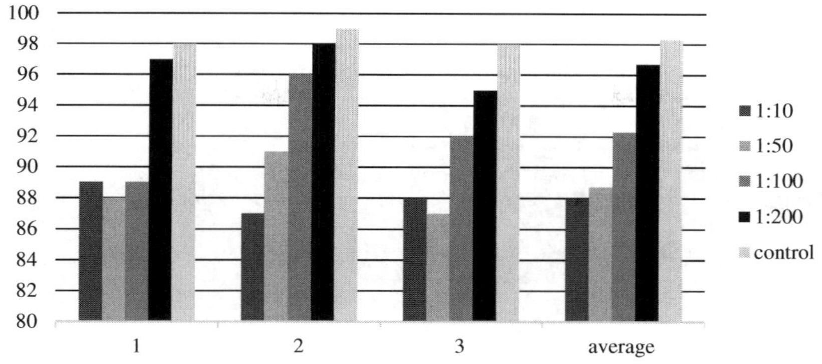

Figure 15.3 Wheat germination (%) at various dilutions of the spent digestate.

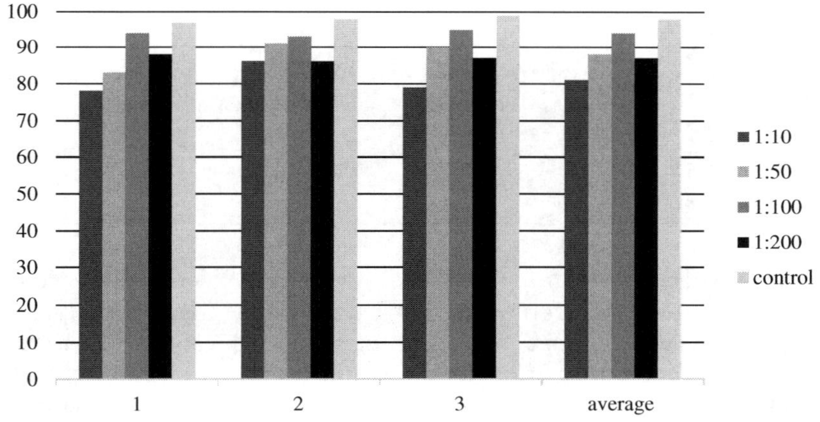

Figure 15.4 Pea germination (%) at various dilutions of the spent digestate.

Seedlings were kept for 5 days in constant light, which was maintained daily for 14 hours (from 6:00 to 20:00). The following indicators were observed:

- growth dynamics, length of aboveground and underground parts of sprouts and their growth;
- the total number of germinated seeds (at the end of the experiment);
- weight of sprouts at the end of the experiment and dry residue.

The results are shown in Table 15.5 and Figure 15.5.

Table 15.5 shows that the average length of the aboveground part increases along with the fertilizer concentration, and the root system is larger compared to the control in all dilutions except 0.09, which can be explained by the achievement of the phytotoxic effect at excessive fertilizer concentration. For each of the studied options, the

Table 15.5 The length of the aboveground and underground parts of the sprouts depending on the digestate concentration

No.	*The mass of the introduced digestate, g*									
	0 (control)		*0.03*		*0.05*		*0.07*		*0.09*	
	Root length, cm	*Stem length, cm*	*Root length, cm*	*Stem length, cm*	*Root length, cm*	*Stem length, cm*	*Root length, cm*	*Stem length, cm*	*Root length, cm*	*Stem length, cm*
1	2	3	4	5	6	7	8	9	10	11
1	15	11	12	9	13	9	12.5	12	8	14
2	12	10	8.5	10	8	10.5	10	10.5	9	14
3	14	11	13.5	12.5	10	12.5	10	9	8	11
4	13	11	7.5	5.5	14	14	7	8	10	14
5	11	9	3	2	15	4.5	5	4	6	9
6	13	9.5	9	7	11	8	9	11.5	5	8
7	7	3	11	8.5	14	14	10.5	12	8	12
8	9.5	10.5	12	12.5	8	8.5	17	11	4	2
9	7	2.5	16	13.5	12	13	6	6	–	–
10	7.5	3.5	5.5	2	7	5	4	5	–	–
11	11.5	9	9.8	7	7.5	4	10	13	–	–
12	7.5	8.5	10.2	9.5	3.5	3	12	15	–	–
13	5	3	9.4	10	7	6	14	13	–	–
14	10	11	9	6.5	10	8.6	12	11	–	–
15	7	6.5	10.6	8.2	–	–	11	13.5	–	–
16	7	3.5	–	–	–	–	–	–	–	–
17	9	1.5	–	–	–	–	–	–	–	–
18	2	1.5	–	–	–	–	–	–	–	–

average length of the aboveground and root systems was calculated (Table 15.6), where m is the arithmetic mean error, which was determined by the formula (15.1):

As can be seen from Table 15.5, the value of $t \leq 3$, which indicates that the results of the study of biometric parameters of seedlings, differs from the control experiment. That is, the intensity of growth processes in plants grown with digestate is better than in the control solution. A negative phytotoxic effect indicates a beneficial effect of biofertilizer on the development of sprouts. In the case of a concentration of 0.09 g, it is possible to achieve a critical amount of digestate. At the end of the experiment, the plants were carefully removed from Petri dishes, whereas the length of the root and stem system of seedlings was measured (Figure 15.6), and the dry weight of seedlings after drying by weight was determined using Ohaus PA 210 analytical scales.

Five medium-sized plants from each Petri dish were weighed on an Ohaus PA 210 analytical scales to determine the wet weight, and after drying, the dry weight was determined. The results of statistical processing of measurement data are shown in Figure 15.7 and Table 15.7.

As can be seen from Figure 15.7, the values of wet and dry mass of seedlings in the Petri dishes with added digestate exceed the following values compared to the control.

The studies also determined the most favorable concentration of spent substrate obtained after BGA biomethanogenesis, namely, 0.07 g. Germination was determined in the percentage of germinated from 20 seeds compared to the control in three

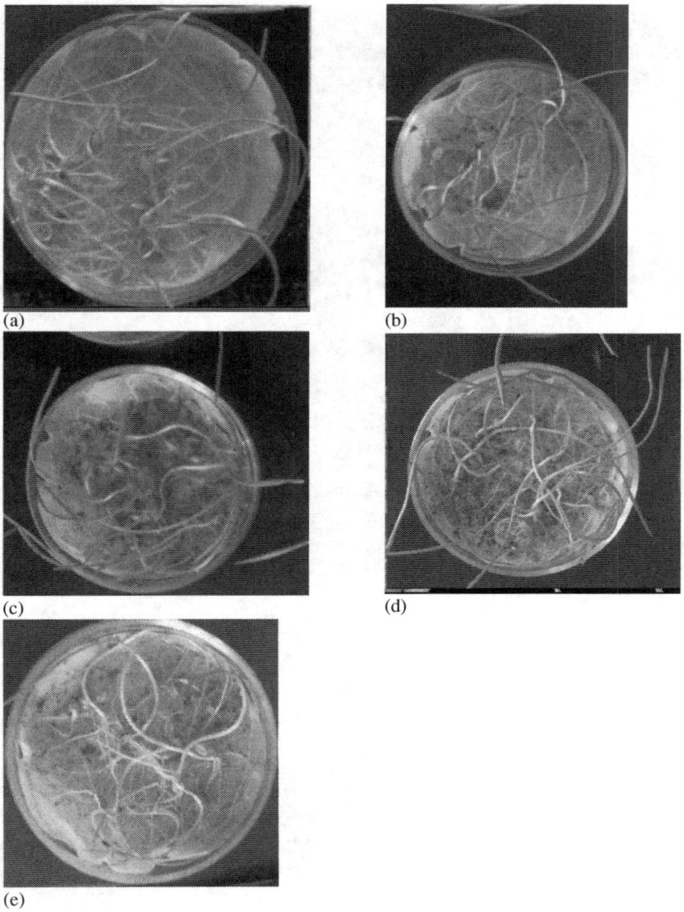

(a) (b) (c) (d) (e)

Figure 15.5 General view of wheat germ in aqueous solutions of digestate (a) control; (b) in a solution of 0.3 g of digestate; (c) 0.05 g of digestate; (d) 0.07 g of digestate; and (e) 0.09 g of digestate.

Table 15.6 The results of mathematical processing of the growth test data

Index	*The mass of the introduced digestate, g*									
	0 (control)		*0.03*		*0.05*		*0.07*		*0.09*	
	Root length, cm	*Stem length, cm*	*Root length, cm*	*Stem length, cm*	*Root length, cm*	*Stem length, cm*	*Root length, cm*	*Stem length, cm*	*Root length, cm*	*Stem length, cm*
X	9.33	6.97	9.80	8.25	10.00	8.61	10.00	10.30	7.25	10.50
$б^2$	11.58	14.07	9.8	11.7	11.19	15.50	11.96	10.60	4.214	17.14
M	0.80	0.88	0.99	1.08	0.93	1.09	0.89	0.84	0.72	1.46
T	–	–	−0.36	−0.91	−0.54	−1.16	−0.55	−2.72	−1.92	−2.06

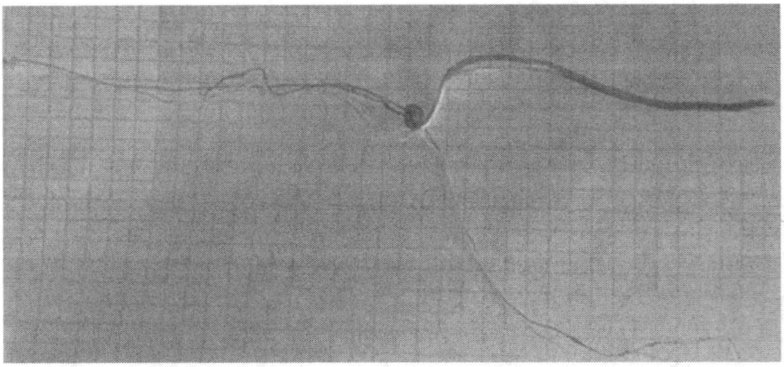

Figure 15.6 Five-day seedling of *Triticum aestivum* L. test culture seeds.

Table 15.7 The average values of wet and dry weights of plant seedlings in one container at different concentrations of the introduced digestate

Index	*The mass of the introduced digestate, g*				
	Control (0)	*0.03*	*0.05*	*0.07*	*0.09*
Wet weight, g	0.282	0.466	0.45	0.46	0.427
Dry weight, g	0.204	0.326	0.302	0.322	0.302

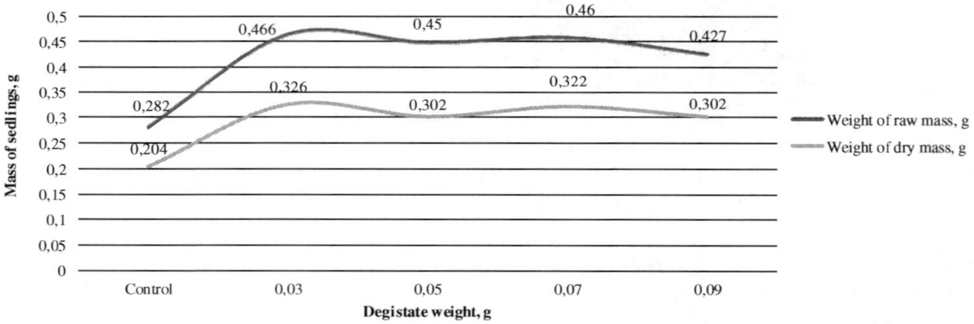

Figure 15.7 The average values of wet and dry weight of *Triticum aestivum* L. seedlings in one container at different concentrations of the introduced digestate.

replicates. During this period, the temperature was +25°C and pH = 6.0. The results are shown in Figure 15.8.

Despite the intensity of wheat germination in the prepared solutions, in the further development, the samples grown with digestate showed better results than in the control, due to the beneficial effect of the spent cyanobacterial substrate after biomethanogenesis on the plant development.

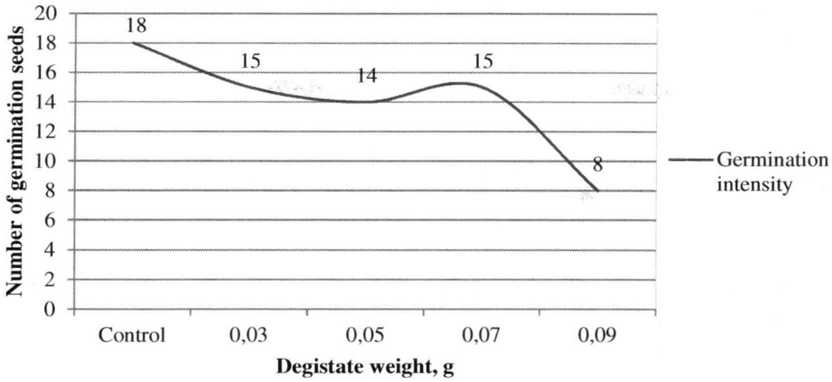

Figure 15.8 Intensity of wheat germination depending on the amount of digestate.

The experimental studies also revealed the possibility of using digestate as a biofertilizer based on a mixture of excess biomass of cyanobacteria from "bloom" spots and activated sludge at the treatment facilities of the municipal utility company "Kremenchukvodokanal". The corresponding components were mixed in a digester before the process of methanogenesis in a ratio of 1:1 by volume.

15.4 CONCLUSIONS

1. For the first time, the effect of specific concentrations of biofertilizer from spent cyanobacterial substrate on test objects – *Triticum aestivum* L. and *Pisum sativum* L. – was investigated; for the first time, the toxicity of the digestate mixture based on biomass of cyanobacteria and activated sludge of municipal treatment facilities was determined; practical ways of utilization of digestate, including its application in agriculture and forestry, were also offered.
2. The biochemical composition of the spent cyanobacterial substrate, the main components of which are macro- and microelements, was studied. The majority of biomass is represented by organic substances; the average mineral content is 4%–6%, which is a prerequisite for the use of digestate as a biofertilizer.
3. The existing methods of the growth test and determination of the phytotoxic effect, as well as preliminary studies on the use of BGA as a biofertilizer, were analyzed. The growth test was performed according to the method of Gorova A. I. and the influence on the growth parameters of the test object *Triticum aestivum* L. and different concentrations of digestate was determined, namely, 0.03, 0.05, 0.07 and 0.09 g, of which the best indicators detected in a dilution of 0.07 g, and when diluted with more than 0.09 g of spent substrate may occur phytotoxic effect.
4. The organic matter of the digestate does not contain harmful substances and has no toxic effect, which was confirmed by complex studies using biotesting methods, and in a number of growth tests the results exceeded the control values. The analysis of the results from the studies on pea and wheat germination under the action of substrates with different concentrations revealed that the germination of wheat

at substrate dilutions of 1:100 and 1:200 is the highest, in turn, in peas this dilution is 1:50 and 1:100. For use as an organic fertilizer, the dilution of the spent substrate of 1:100 is optimal, both for wheat and peas.

5. The respective dilutions were also subjected to a biotesting procedure according to the main method of determining the national water quality standard, which provides for the use of crustaceans as test objects. It also did not show any acute or chronic toxic effects of dilute spent organic matter on the living organisms used in the test.
6. The results of the studies on the positive effect of certain concentrations of spent after biomethanogenesis digestate based on the biomass of cyanobacteria and other organic substrates of the phyto- and zoogenic origin on the development of cultivated plants can be used in agriculture and forestry. The biotesting indicators also allow determining the optimal concentrations of the studied organo-mineral fertilizer in the agro-industrial complex.

REFERENCES

Abiru, T., Mitsugi, F., Ikegami, T., Ebihara, K., Aoqui, S.-I. & Nagahama, K. 2015. Environmental application of electrical discharge for ozone treatment of soil. *Informatyka, Automatyka, Pomiary w Gospodarce i Ochronie Środowiska* 5(4): 42-44. doi:10.5604/20830157.1176573

Agadzhanyan, N.A. Ushakov, I.B. & Torshin, V.I. 1997. Ecologiya cheloveka: Slovar-spravochnik/Avt.-sost. i dr.; Pod obshch. red. N.A. Agadzhanyan. MMP "Ecotsentr", izdatelskaya firma "KRUK", 208 p.

Azarov, O.D., Murashchenko, O.G., Chernyak, O.I., Smolarz, A. & Kashaganova, G. 2015. Method of glitch reduction in DAC with weight redundancy. *Proceedings of the SPIE* 9816: 98161T.

Digtiar, S. 2016. Qualitative and quantitative characteristics of biogas of cyanea organic mass. *Environmental Problems* 1(2): 149–153.

DSTU ISO 5667-2:1991. Yakist vody. Vidbyrannya prob. Chastyna 2. Nastanovy shchodo metodiv vidbora prob.

Gorova, A. & Kulina, S. 2008. Otsinka toxychnosti gruntiv Chervonohradskogo hirnychopromyslovogo rayonu za dopomogoyu rostovogo testu. *Visn. Lviv. un-tu. Seriya biologichna* 48: 189–194.

Lakin, G.F. 1990. *Biometriya*. Vysshaya shkola, 352 p.

Lugovoy, A.V., Yelizarov, O.I., Nykyforov, V.V. & Digtiar, S.V. 2007. Sposib otrymannya biogazu z synyozelenyh vodorostey. Patent na korysnu model 24106 vid 25 chervnya 2007 roku. – *K.*, Bulletin No. 9.

Malovanyy, M., Nikiforov, V., Kharlamova, O. & Synelnikov, O. 2016. Production of renewable energy resources via complex treatment of cyanobacteria biomass. *Chemistry & Chemical Technology* 10(2): 251–254.

Natsionalnyi standart Ukrayiny: Yakist vody. (ISO 6341:1996, MOD) DSTU 4173:2003.

Nykyforov, V.V. & Kozlovs'ka, T.F. 2002. Khimiko-biologicheskiye prichiny uhudsheniya kachestva prirodnoy vody // Visnyk KDPU. *Kremenchuk* 6(17): 82–85.

Nykyforov, V.V., Kozlovs'ka, T.F. & Digtiar, S.V. 2008. Khimicheskaya biologiya methanogenesa sine-zelyonyh vodorosley i polozhytelnye effecty ih utilizatsii. Ecologychna bezpeka. *Kremenchuk* 2 (2): 83–91.

Nykyforov, V.V., Kozlovs'ka, T.F., Novokhatko, O.V. & Digtiar, S.V. 2018. On additional possibilities of using the blue-green algae substrate and digestate. *Water Supply and Wastewater Disposal*:

207–220. https://www.researchgate.net/profile/Nataliya-Bernatska/publication/335001174_Water_Supply_and_Wastewater_Disposal_Monografie_-Politechnika_Lubelska_WATER_SUPPLY_AND_WASTEWATER_DISPOSAL_5/links/5d498e1392851cd046a6959e/Water-Supply-and-Wastewater-Disposal-Monografie-Politechnika-Lubelska-WATER-SUPPLY-AND-WASTEWATER-DISPOSAL-5.pdf#page=207

Nykyforov, V.V., Malovanyy, V., Kozlovs'ka, T., Novokhatko, O. & Digtiar, S. 2016. The biotechnological ways of blue-green algae complex processing. *Eastern-European Journal of Enterprise Technologies* 5(83): 11–18.

Nykyforov, V.V., Yelizarov, M.O., Pasenko, A.V., Digtiar S.V. & Shlyk, S.V. 2016. Sposib vyrobnytstva metanu ta dobryva. Patent na korysnu model 104743 vid 10 lyutogo 2016 roku. – *K.*, Bulletin No. 3.

Nykyforov, V.V., Yelizarov, O.I., Lugovoy, A.V., Kozlovs'ka, T.F. & Yelizarov, M.O. 2011. Nature protection and energy-resource saving technology of green-blue algae utilization in Dnieper reservoirs. *Transactions of Kremenchuk Mykhailo Ostrohradskyi National University – Kremenchuk: KrNU* 1(66): 115–117.

Osadchuk, V.S. & Osadchuk, A.V. 2011. The magneticreactive effect in transistors for construction transducers of magnetic field. *Electronics and Electrical Engineering – Kaunas: Technologija* 3(109): 119–122.

Pavlov, S.V., Kozhemiako, V.P., Kolesnik, P.F., Kozlovs'ka, T.I. & Dumenko, V.P. 2010. *Physical Principles of Biomedical Optics: Monograph.* Vinnytsya: VNTU: 152 p.

Valtchev, V.S., Teixeira, J.P. & Pavlov, S. 2016. Energy harvesting: An interesting topic for education programs in engineering specialities. In Proceedings of the tenth international scientific-practical conference (IES-2016), 149–156.

Vasilevskyi, O.M., Yakovlev, M.Y. & Kulakov, P.I. 2017. Spectral method to evaluate the uncertainty of dynamic measurements. *Technical Electrodynamics* 4: 72–78.

Zagirnyak, M., Chornyi, O., Nykyforov, V., Sakun, O. & Panchenko, K. 2016. Experimental research of electromechanical and biological systems compatibility. *Przegląd elektrotechniczny* 1: 128–131.

Zakharenko, M.O., Iaremchuk, O.S., Shevchenko, L.V., Polakovski, V.M., Mikhalska, V.M., Maluga, L.V., & Ivanova, O.V.. 2017. Gigiyena ta biofermentatsiya pobichnyh produktiv tvarynnytstva [monograf] – K.: "Tsentr uchbovoyi literatury", 536 p.

Zuman, B.V., Plakushchiy, V.O. & Digtiar, S.V. 2013. Rezultaty doslidzhennya ecologichnoyi sytuatsii na kaskadi vodoskhovyshch richky Dnipro//Novi tehnologii. *Kremenchuk: KUEITU* 1–2(39–40): 106–109.

Chapter 16

The Use of *Microcystis aeruginosa* Biomass to Obtain Fungicidal Drugs

Volodymyr V. Nykyforov, Oksana A. Sakun, Olga V. Novokhatko, and Valeria S. Shendryk
Kremenchuk Mykhailo Ostrohradskyi National University

Katharina Meixner
University of Natural Resources and Life Science

Anastasiia A. Cherepakha
Vinnytsia National Technical University

Zbigniew Omiotek
Lublin University of Technology

Saltanat Kalimoldayeva
Regional Diagnostics Center

Dina Nuradilova
Asfendiyarov Kazakh National Medical University

CONTENTS

16.1 INTRODUCTION

Water and health, and in particular drinking water and health, have been an area of concern to the World Health Organization (WHO) for many years. A major activity of WHO is the development of the guidelines, which present an authoritative assessment of the health risks associated with the exposure to infectious agents and chemicals through water. Such guidelines already exist for drinking water and for

DOI: 10.1201/9781003177593-16

the safe use of wastewater and excreta in agriculture and aquaculture; they are currently being prepared for the recreational uses of water in co-operation with the United Nations Educational, Scientific and Cultural Organization, United Nations Environment Programme and the World Meteorological Organization (Chorus & Bartram 1999).

The WHO supports the development of national and international policies concerning water and health, and assists countries in developing the capacities to establish and maintain healthy water environments, including legal frameworks, institutional structures, and human resources. In 2018, WHO included the emergence and spread of cyanobacteria in the list of world problems with a potentially high level of danger. It is a cosmopolitan, their blood was recorded in more than 80 countries around the world for the period 2010–2018 (Faassen & Lurling 2013).

Blue-green microalgae (*Cyanophyta*) or cyanobacteria (*Cyanobacteria*) synthesize a large number of secondary metabolites. Particular attention is paid to algotoxins among them, since they pose a danger to the life and health of humans and animals. Algotoxins are released during the massive development of microalgae – the so-called "blooming" of eutrophic water bodies. All these prevent using the water for the economic and recreational purposes; a violation and degradation of biohydrocenoses occurs (Malovanyy et al. 2019; Nykyforov et al. 2019).

Microcystins (MCs) are some of the most typical cyanotoxins in fresh water. Their main producers are toxinogenic cyanobacteria of the *Aphanizomenon*, *Anabaena*, *Anabaenopsis*, *Microcystis*, *Oscillatoria*, *Phormidium* genera, and others. From a biochemical point of view, MCs are cyclic heptapeptides with three variable methyl groups (Figure 16.1). MCs are synthesized by large enzyme complexes consisting of nonribosomal polypeptide synthetases.

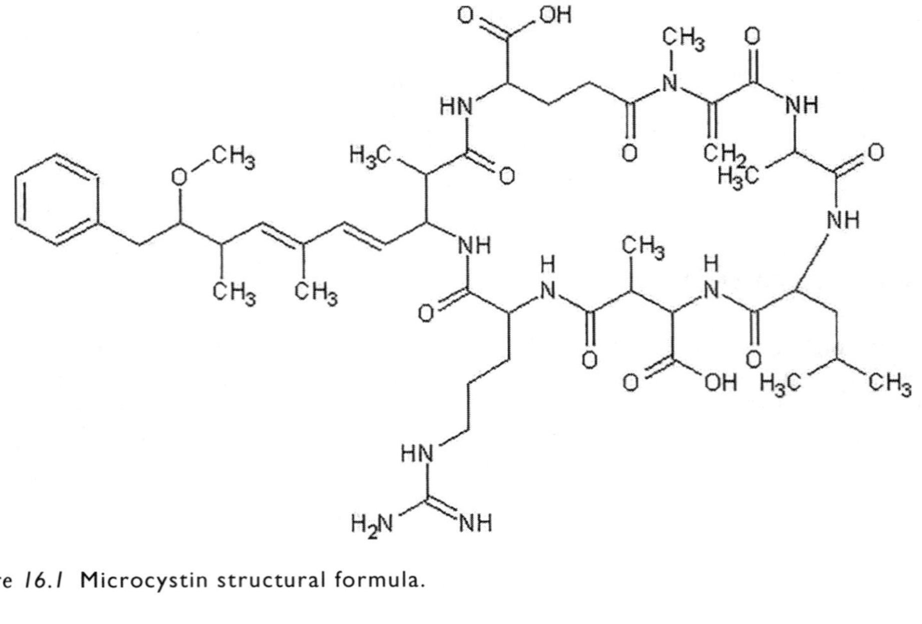

Figure 16.1 Microcystin structural formula.

MCs are well-soluble in water and remain stable up to 7 days. MCs are resistant to hydrolysis or oxidation at pH ≈ 7, and upon boiling they do not break down for several hours. At high temperatures and high or low pH, more than 90% of MCs decomposes in 3 months. The toxicity of MCs is directed to the organs carrying out the transfer of organic ions through the plasmalemma (mainly to the liver).

MCs are influenced by humans when drinking water, as well as through the skin when swimming in water. According to the WHO recommendations, the concentration of MC-LR in drinking water should not exceed 1 μg/dm^3 for single use and 0.1 μg/dm^3 for repeated use. When using water bodies for recreational purposes, the number of cyanobacteria 2×10^7 cells/dm^3 and MCs concentration 2–4 μg/dm^3 is a threat. Human LD_{50} for MCs ranges from 0.05 to 1.20 mg/kg body weight. The clinical symptoms of MCs intoxication include diarrhea, nausea, chills, weakness, and pallor. Death in acute MCs poisoning occurs as a result of severe obstruction of the blood vessels of the liver and extensive hepatic hemorrhage. The dermatological symptoms upon the contact with MCs include blisters on the lips, allergic reactions (contact dermatitis, asthma, hay fever, and conjunctivitis) (Zurawell et al. 2005; Buratti et al. 2013).

In 2009, an analysis based on a valuable polymerase reaction for the presence of genes encoding the synthesis of microcystins was positive for 68% of the studied reservoirs in Ukraine, in 2010 – for 90% of the reservoirs. Using chromatography-mass spectrometry in the phytoplankton of the Dnieper reservoirs, microcystins LR, RR, YR, etc., as well as eruginosins – linear peptides with antithrombin activity, were determined. The analysis showed that the representatives of the *Microcystis* genus are potentially toxic (Belykh et al. 2013; Wójcik et al. 2019).

There are a large number of publications devoted to the biochemical, genetic, physiological, and toxicological studies of MCs. The literature also describes the processes of MCs bioaccumulation in various plant and animal organisms used for food. In our opinion, it is very important to study the presence of MCs congeners in food products, since there are differences in the mechanisms of absorption and how they are excreted from the body (Gutierrez-Praena et al. 2013; Ame et al. 2010; Buratti et al. 2017). Unlike previous studies, this article considers the possibility of using the algotoxins of blue-green algae as biopesticides in general and microcystin as a fungicide in particular. Therefore, our goal is to study the effect of MCs on the development of tomatoes late blight.

16.2 MATERIALS AND METHODS

The concentration of microcystin was determined by Austrian colleagues using chromatography-mass spectrometry in two water samples from the Kremenchug reservoir, taken at different times with an interval of 2 weeks. The samples were taken from *Microcystis aeruginosa* (Kützing) Kützing clusters at a water temperature of +25°C. In order to determine the optical density, a wavelength of $\lambda = 450$ nm was used.

There are several methods for the extraction and cultivation of late blight, differing from each other in the composition of the culture medium, antiseptics, and other sterilization methods (Hamed & Gisi 2012; Petruk et al. 2012). In this work, we used a method for the preparation of isolates from the surface tissues of tomato fruits infected with late blight. Nutrient agar from "Pharmactive LLC" TU U 24.4-37219230-001:2011 (enzyme peptone 10 g/l, microbiological agar 10 g/l, sodium chloride 5 g/l, yeast extract 3 g/l). In order to prepare 1 liter of culture medium, 28 g of dry matter was used. The mixture was boiled for 3 minutes, sterilized, and cooled to 45°C, poured into sterile Petri dishes, and left to solidify. The prepared culture medium has a pH of 7.3 ± 0.2.

Pretreatment of the starting material with an antiseptic is an important step in eliminating contamination from extraneous microflora. Potassium permanganate and ethyl alcohol were selected from the list of available disinfectants (Figure 16.2). The first and second stages of processing is the alternate soaking of isolates in the solutions of potassium permanganate and ethyl alcohol every 5 minutes. The third step is washing the starting material for 1 minute with distilled water. After disinfection, the starting material is placed in a sterile Petri dish. Before processing, the raw material must be cut into pieces of approximately 5 by 5 mm.

Then, three successive reseeding of the thus obtained pure culture of *Phytophthora infestans* (Mont.) De Bary was carried out every 3 days. Prior to transferring the isolates to a fresh nutrient medium, the crops were treated with a 1 M ampicillin solution for additional decontamination. Given the growth rate of the culture, the isolates were transferred to a fresh nutrient medium in an exponential phase.

Next, optical microscopy of phytophthora colonies was carried out using the "Micromed" trinocular digital microscope. The heterogeneity of morphological characters was established (Figure 16.3), which manifests itself in the difference in the structure, size, topography, and growth rate of the colonies. There was also a different ability to form surface mycelium. The diameter of the colonies ranged from 2 to 5 mm. Some colonies were characterized by a weak development of aerial mycelium, while others, on the contrary, were quite strong. The surface of the mycelium was cotton-like or arachnoid.

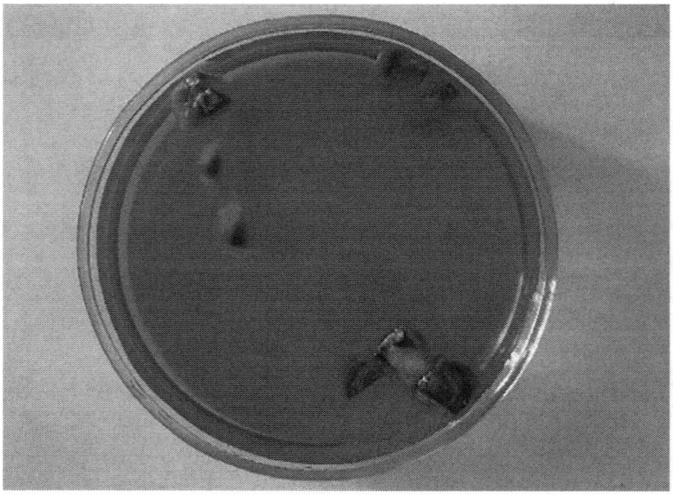

Figure 16.2 Decontamination of isolates from the surface tissues of tomato fruits.

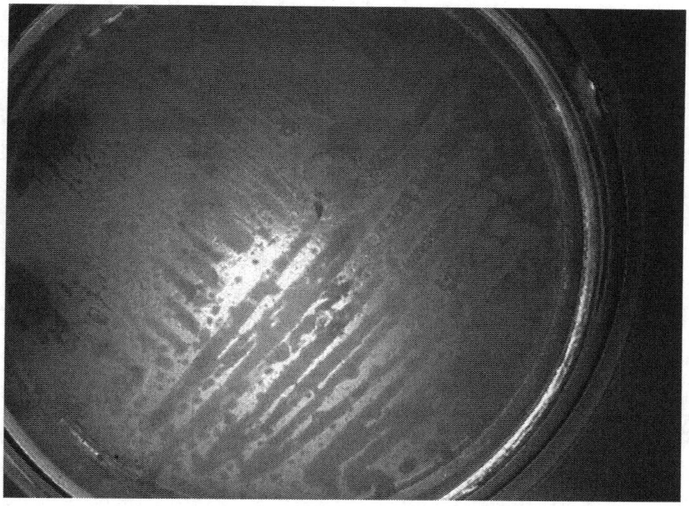

Figure 16.3 *Phytophthora infestans* (Mont.) de Bary colonies.

16.3 RESULTS AND DISCUSSION

The aqueous suspension of *Microcystis aeruginosa* (Kützing) Kützing cells has a relative density of 0.96–0.98 and contains 1–2 g of dry matter/1 dm^3. The concentration of microcystin (cyclopeptide algotoxin) in suspension varies between 3.25 and 3.48 µg/dm^3 (Table 16.1 and Figure 16.4).

The study of the quantitative dynamics of *Phytophthora infestans* (Mont.) De Bary cultures under the influence of microcystin in a suspension of blue-green algae was carried out simultaneously with the analysis of the univariate effect of antibiotics or

Table 16.1 The microcystin concentration in an aqueous suspension of *Microcystis aeruginosa* (Kützing) Kützing cells

Sample	OD_1	OD_2	*OD Average*	Mcy_1 mg/dm^3	Mcy_2 mg/dm^3	*Average* mg/dm^3	*Standard deviation*	B/B_0, %	*CV*, %
Negative control	0.772	0.722	0.747	0.01	0.02	0.017	0.01	100	71.8
Standard control 0.1	0.612	0.641	0.627	0.05	0.04	0.045	0.06	84	55.4
Standard control 0.5	0.295	0.284	0.289	0.63	0.69	0.663	0.16	39	32.6
Standard control 1	0.181	0.195	0.188	1.58	1.41	1.498	0.50	25	49.8
Standard control 2	0.140	0.129	0.135	2.20	2.40	2.300	0.30	18	15.0
Standard control 5	0.10	0.107	0.104	3.03	2.86	2.948	2.05	14	41.0
Dnieper 1A	0.09	0.108	0.090	3.28	2.84	3.249	0.31	12	–
Dnieper 1B	0.079	0.086	0.083	3.59	3.39	3.484	0.18	11	–

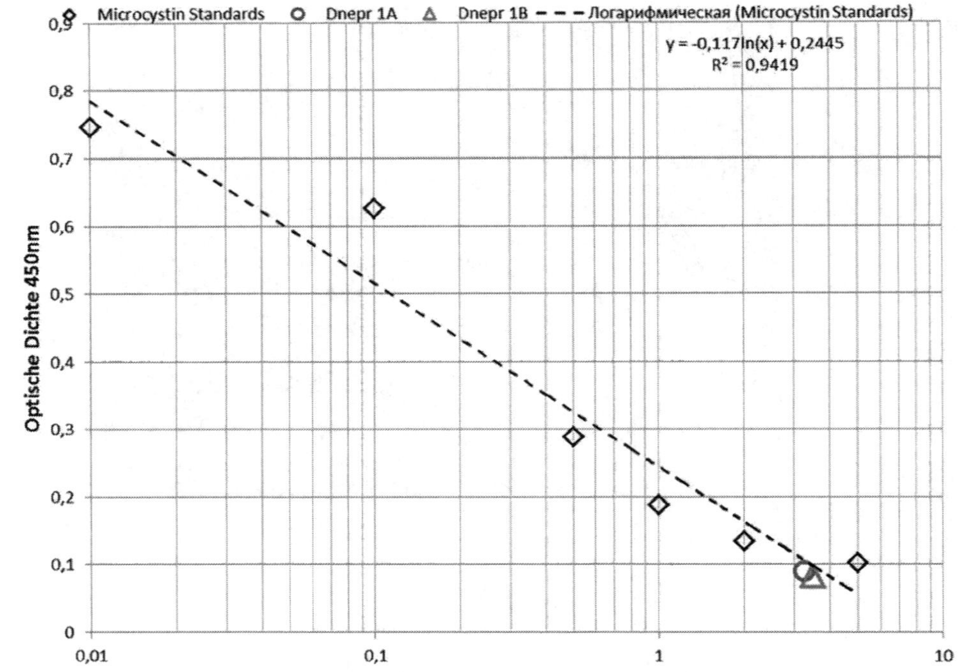

Figure 16.4 Microcystin concentration in the Dnieper water (*OX*, μg/dm^3) at optical density (*OY*) for $\lambda = 450$ nm.

fungicides on the culture without extraneous factors. In total, four experimental lines were drawn. The starting material for isolates was further processed only with potassium permanganate, since this antiseptic gives better results compared to ethanol, after which the colonies had weakened growth. Efficiency was assessed visually by the number of colonies that formed on the nutrient medium during 5 days of growth.

Experimental line No. 1 is a control. Experimental line No. 2 provides for the treatment of phytophthora colonies with a solution of ampicillin (2 g per 1 liter of phosphate buffer) in an amount of 1 ml per one Petri dish. In experimental line No. 3, the phytophthora colonies were treated with a solution of fungicide "Quadris" (2 g/1 l of bidistillate) in an amount of 1 ml per Petri dish. Experimental line No. 4 involves adding 1 ml of a suspension of blue-green algae to a Petri dish.

The population of *Phytophthora infestans* (Mont.) De Bary *in vitro* is characterized by a very dynamic development: the number of colonies increases exponentially. In order to build a graph of population growth, we observed and counted the number of colonies daily at the same time. It has been reported that the full cycle of development and degradation of the phytophthora population can last an average of 20 days. In some cases, the duration of the phases can vary by several days.

The growth phases of the *Phytophthora* population *in vitro* in experimental line No. 1 (control) are shown in Figure 16.5. The lag phase is observed on days 1 and 2. From the beginning of culture inoculation, 24 colonies formed. During the day (on the fourth day), the number of colonies more than doubled (51 colonies). From the fourth

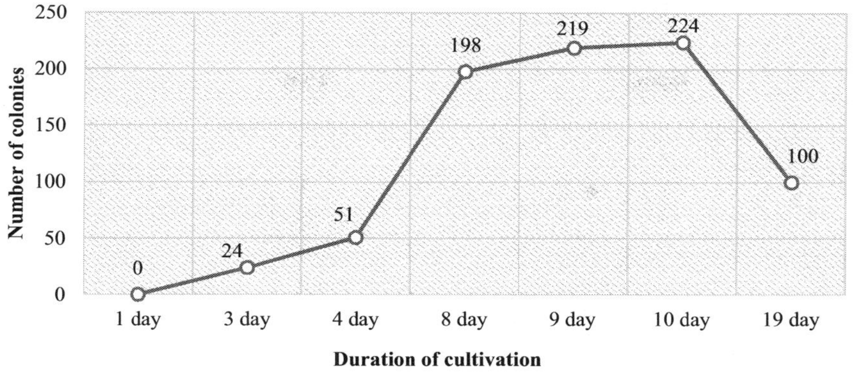

Figure 16.5 Growth phases of the *Phytophthora* population *in vitro* (experimental line No. 1: control).

to the eighth days, an exponential growth phase is observed (198 colonies), which ends as quickly as it begins. On the ninth day, the culture goes into a phase of slow growth. Then, as a rule, on the 10th day, the stationary phase sets in – for a while the population freezes – an increase of five colonies. Further, a uniform withering of the culture is observed. On the 19th day, the number of viable colonies decreases by more than two times. The experiment ends on the 20th day.

In experimental line Nos. 2–4, various growth inhibitors of *Phytophthora infestans* (Mont.) De Bary were added to the nutrient medium. For the purity of the experiment, they were introduced on the second day after the start of the exponential phase (on the sixth *in vitro* cultivation). On the basis of the results of observing the inhibition of *Phytophthora* culture with ampicillin, a growth curve was constructed (Figure 16.6). At the beginning of the exponential phase, 34 colonies grew. After 2 days, there were 152 colonies. The day after the introduction of ampicillin, the growth of colonies is sharply reduced, and after 2 days, the number of colonies decreased by 21.7%.

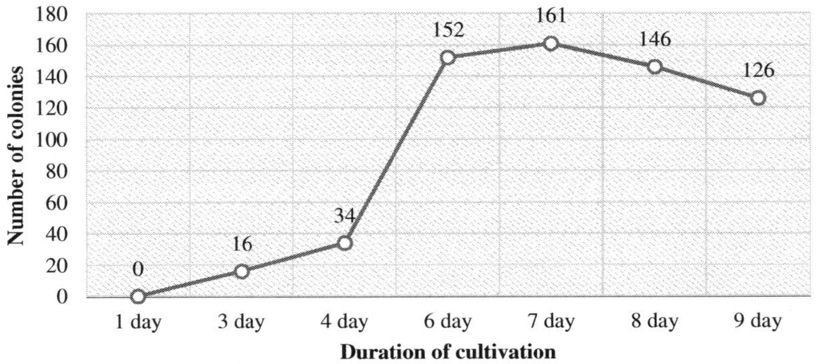

Figure 16.6 Growth phases of the *Phytophthora* population *in vitro* (experimental line No. 2: ampicillin).

A graph was also plotted showing the inhibition of Phytophthora culture by the "Quadris" fungicide in experiment No. 3 (Figure 16.7). At the beginning of the exponential phase (day 4 of observation), there were 47 colonies in the culture. After 2 days (6 days), the growth of the colonies was 123. At this time, the culture was treated with fungicide. Afterwards, the growth of the colonies stopped, and after 2 days, the number of colonies decreased by 34.1%.

In our opinion, special attention should be paid to experimental line No. 4, since it is a key in terms of using microcystins for the treatment and prevention of cultivated plants late blight, as well as the effectiveness of using of blue-green algae suspension (90% *Microcystis aeruginosa*)(Kützing) as a biopesticide with fungicidal properties. At the beginning of the exponential phase (4 days), the population numbered 40 colonies. Two days later, their number increased sharply by 159 colonies (Figure 16.8, Table 16.2). On the sixth day, the culture was treated with 1 ml of blue-green algae suspension (70–100 g/m^3). A day later (on the seventh day), a reduction of 13 colonies was recorded, on the eighth day, another 16, and on the ninth day, another 15 colonies. Thus, the

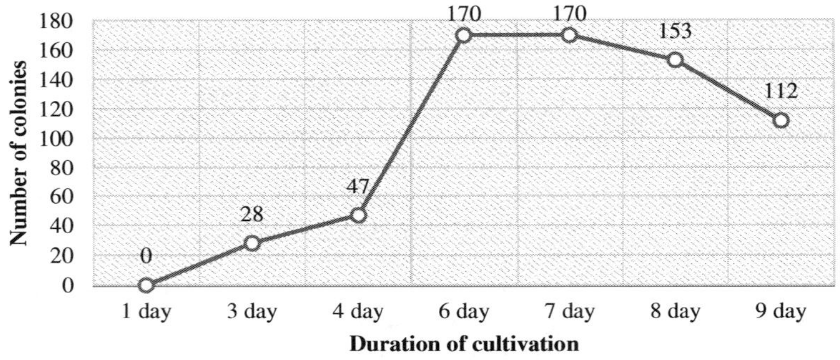

Figure 16.7 Growth phases of the *Phytophthora* population *in vitro* (experimental line No. 3: "Quadris" fungicide).

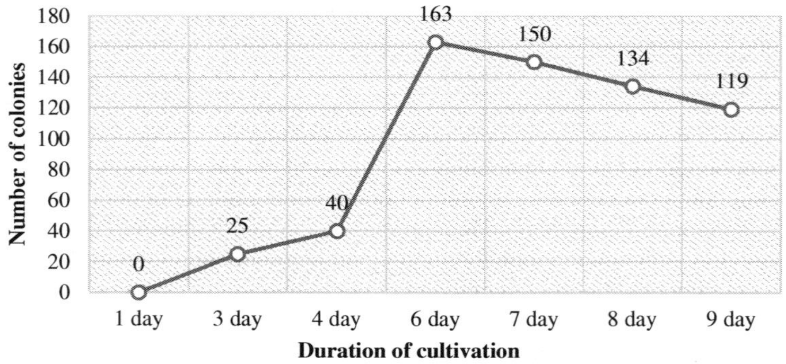

Figure 16.8 Growth phases of the *in vitro Phytophthora* population (experimental line No. 4: blue-green algae suspension).

Table 16.2 The colonies growth dynamics of *Phytophthora infestans* (Mont.) De Bary in experimental lines using different growth inhibitors

Δ Number of colonies		*After treatment with an inhibitor*			*Total, %*
On day 4	*On day 6*	*In 1 day*	*In 2 day*	*In 3 day*	
34	118	9	−15	−20	−21.7
47	123	0	−17	−41	−34.1
40	123	−13	−16	−15	−20.7

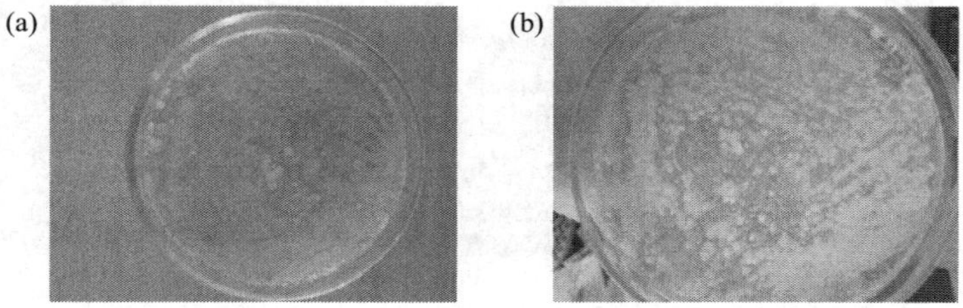

Figure 16.9 Degradation of *Phytophthora infestans* (Mont.) De Bary colonies after suspension treatment: (a) 6 days and (b) 3 days after injection of blue-green algae.

decrease in the population of *Phytophthora infestans* (Mont.) De Bary *in vitro* as a result of processing its culture was 20.7%, which is visible in Figure 16.9.

At the next stage of research, the fungicidal efficacy of cyanobacteria was determined depending on the method of processing the affected tomatoes (*Solanum lycopersicum* L). Three experimental lines used a hundred plants of the "Room" variety. Conventionally, four pathogenic stages of late blight were identified: (1) the lower tier of the leaves is affected; (2) two tiers of leaves and a stem between them are affected; (3) almost all or all vegetative organs are affected; and (4) all vegetative and generative organs are affected. For 2 weeks, the plants were daily treated with the blue-green algae suspension in two ways: by spraying the ground part and watering the soil, but in equal amounts. The control line of tomatoes was watered and sprayed with water.

Thus, in the course of laboratory studies, it was found that the efficiency of spraying tomatoes with the blue-green algae suspension (75%) at the first two stages of pathogenesis (Figure 16.10) is three times higher compared to irrigation (25%). It was also proven that in the last stages of late blight, none of the methods used is effective.

Along with the results obtained, the question arises of the safety of eating tomatoes after treatment with a suspension of toxic cyanobacteria. Therefore, at the final stage of the study, the toxicity levels of the processed tomatoes were determined using the bioassay method. As a test object, the species of land (pulmonary) gastropods was used – giant Achatina (*Achatina fulica* Ferussac). The choice is due to several reasons: this snail is very sensitive to the contamination of food and habitat, it quickly and adequately responds to negative factors and is also omnivorous and unpretentious in content (Chesnokova 2008; Gutierrez-Praena et al. 2013; Drobac 2017).

During the biotesting, it was found that plants that were watered with a suspension did not cause violent reactions in the test object. In contrast, the snails reacted

Figure 16.10 Fungicidal efficacy of cyanobacteria: (a) control (second stage of late blight) and (b) restoration of *Solanum lycopersicum* L. on day 14 after spraying with suspension.

Figure 16.11 Closure of *Achatina fulica* Ferussac shells with protective plates as a result of the injection of crushed tomato leaves sprayed with a cyanobacteria suspension for 14 days.

to the plants that were sprayed with a sharp decrease in activity and clogging of shells (Figure 16.11). A similar test object reaction was observed in 12 out of 31 individuals tested by *Lactuca sativa* L, and in 8 out of 40 individuals tested by *Capsicum annuum* L (Kvaternyuk et al. 2017; Petruk et al. 2015; Wójcik & Smolarz 2017).

16.4 CONCLUSIONS

The aqueous suspension of *Microcystis aeruginosa* (Kützing) Kützing cells has a relative density of 0.96–0.98 and contains 1–2 g of dry matter/1 dm^3. The concentration of microcystin (cyclopeptide algotoxin) in the suspension varies between 3.25 and 3.48 $\mu g/dm^3$.

The population of *Phytophthora infestans* (Mont.) De Bary *in vitro* is characterized by a very dynamic development: the number of colonies increases exponentially. In order to build a graph of population growth, we observed and counted the number of colonies daily at the same time. It was reported that the full cycle of development and degradation of the phytophthora population can last an average of 20 days.

The lag phase for experimental line No. 1 (control) is observed on days 1 and 2. From the beginning of culture inoculation, 24 colonies formed. During the day (on the fourth day), the number of colonies more than doubled (51 colonies). From the fourth to the eighth days, an exponential growth phase is observed (198 colonies). On the ninth day, the culture goes into a phase of slow growth. Then, as a rule, on the 10th day, the stationary phase sets in – for a while the population freezes – an increase of five colonies.

In experimental lines Nos. 2–4, various growth inhibitors of *Phytophthora infestans* (Mont.) De Bary were added to the nutrient medium. For the purity of the experiment, they were introduced on the second day after the start of the exponential phase (on the sixth *in vitro* cultivation).

On the basis of the results of observing the inhibition of *Phytophthora* culture with ampicillin, a growth curve was constructed. At the beginning of the exponential phase, 34 colonies grew. After 2 days, there were 152 colonies. The day after the introduction of ampicillin, the growth of colonies is sharply reduced, and after two days, the number of colonies decreased by 21.7%.

At the beginning of the exponential phase of experimental line No. 3 (day 4 of observation), there were 47 colonies in the culture. After 2 days (6 days), the growth of the colonies was 123. At this time, the culture was treated with fungicide. Afterwards, the growth of the colonies stopped, and after 2 days, the number of colonies decreased by 34.1%.

The experimental line No. 4 is key in terms of using microcystins for the treatment and prevention of cultivated plants late blight, as well as the effectiveness of using the blue-green algae suspension as a biopesticide with fungicidal properties.

At the beginning of the exponential phase (4 days), the population numbered 40 colonies. Two days later, their number increased sharply by 159 colonies. On day 6, the culture was treated with 1 ml of the blue-green algae suspension. A day later (on the seventh day), a reduction of 13 colonies was recorded, on the eighth day, another 16, and on the ninth day, another 15 colonies. Thus, the decrease in the population of *Phytophthora infestans* (Mont.) De Bary *in vitro* as a result of processing its culture was 20.7%.

It was found that the efficiency of spraying tomatoes with the blue-green algae suspension (75%) at the first two stages of pathogenesis is three times higher compared to irrigation (25%). It was also proven that in the last stages of late blight, none of the methods used is effective.

During the biotesting (as a test object, the species of land gastropods was used – *Achatina fulica* Ferussac), it was found that the plants that were watered with a suspension did not cause violent reactions in the test object. In contrast, snails reacted to the plants that were sprayed with a sharp decrease in activity and clogging of shells.

REFERENCES

Ame, M., Galanti, L. & Menone, M. 2010. Microcystin-LR, -RR, -YR and -LA in water samples and fishes from a shallow lake in Argentina. *Harmful Algae* 9: 66–73.

Belykh, O. I., Gladkikh, A. S., Sorokovikova, E. G., Tikhonova, I. V., Potapov, S. A., & Fedorova, G. A. 2013. Microcystin-producing cyanobacteria in reservoir of Russia, Belarus and Ukraine. *Chemistry in the Interests of Sustainable Development* 21(4): 363–378.

Buratti, F. et al. 2013. The conjugation of microcystin-RR by human recombinant GSTs and hepatic. *Toxicology Letters* 219: 231–238.

Buratti, F. et al. 2017. Cyanotoxins: producing organisms, occurrence, toxicity, mechanism of action and human health toxicological risk evaluation, *Archives of Toxicology* 91: 1049–1130.

Chesnokova, S. M. 2008. *Biologycheskije metody otsenki kachestva obektov okruzhaiushchei sredy. Chast 2. Metody biotestirovanija: uchebn. posob.* Vladymyr: VlHU: 92.

Chorus, I. & Bartram, J. eds. 1999. *Toxic Cyanobacteria in Water. A Guide to Their Public Health Consequences, Monitoring, and Management.* London and New York: Routledge: 416. https://www.who.int/water_sanitation_health/resourcesquality/toxcyanbegin.pdf

Drobac, D. 2017. Microcystin accumulation and potential effects on antioxidant capacity of leaves and fruits of Capsicum annuum. *Journal of Environmental Science and Health, Part A* 80: 145–154.

Faassen, E. & Lurling, M. 2013. Occurrence of the Microcystins MC-LW and MC-LF in Dutch surface waters and their contribution to total microcystin toxicity. *Marine Drugs* 11: 2643–2654.

Gutierrez-Praena, D. et al. 2013. Presence and bioaccumulation of microcystins and cylindrospermopsin in food and the effectiveness of some cooking techniques at decreasing their concentrations: a review. *Food and Chemical Toxicology* 53: 139–152.

Hamed, B. & Gisi, U. 2012. Generation of pathogenic F1 progeny from crosses of Phytophthora infestans isolates differing in ploidy. *Plant Pathology* 19: 1365–3059.

Kvaternyuk, S. et al. 2017. Increasing the accuracy of multispectral television measurements of phytoplankton parameters in aqueous media. *SGEM2017 Vienna GREEN Conference Proceedings*, 17(33): 219–225, Sofia, Bulgaria.

Malovanyy, M., Zhuk, V., Nykyforov, V., Bordun, I., Balandiukh, I. & Leskiv G. 2019. Experimental investigation of *Microcystis aeruginosa* cyanobacteria thickening to obtain a biomass for the energy production. *Journal of Water and Land Development* 43(1): 113–119.

Nykyforov, V., Malovanyy, M., Aftanaziv, I., Shevchuk, L. & Strutynska L. 2019. Developing a technology for treating blue-green algae biomass using vibro-resonance cavitators. *Naukovyi Visnyk Natsionalnoho Hirnychoho Universytetu* 6: 181–188.

Petruk, V. et al. 2012. The spectral polarimetric control of phytoplankton in photobioreactor of the wastewater treatment. *Proceedings of the SPIE* 8698: 86980H.

Petruk, V. et al. 2015. The method of multispectral image processing of phytoplankton for environmental control of water pollution. *Proceedings of the SPIE* 9816: 98161N.

Wójcik, W. & Smolarz, A. 2017. *Information Technology in Medical Diagnostic.* CRC Press, London.

Wójcik, W., Pavlov, S. & Kalimoldayev, M. 2019. *Information Technology in Medical Diagnostics II.* Taylor & Francis Group, CRC Press, London.

Zurawell, R. et al. 2005. Hepatotoxic cyanobacteria: a review of the biological importance of microcystins in freshwater environments, *Journal of Toxicology and Environmental Health* 8: 1–37.

Chapter 17

Experimental Research of Engine Characteristics Working on the Mixtures of Biodiesel Fuels Obtained from Algae

Sviatoslav Kryshtopa, Liudmyla Kryshtopa, and Myroslav Panchuk
Ivano-Frankivsk National Technical University of Oil and Gas

Victor Bilichenko
Vinnytsia National Technical University

Andrzej Smolarz
Lublin University of Technology

Aliya Kalizhanova
Institute of Information and Computational Technologies CS MES RK
Kazakhstan University of Power Engineering and Telecommunications

Sandugash Orazalieva
Almaty University of Power Engineering and Telecommunications (AUPET)

CONTENTS

17.1 INTRODUCTION

It should be noted that cheap and light oil is suitable for its exhaustion and exploration of new deposits and extraction with subsequent processing of hard-to-reach high-grade and high-viscosity grades of oil requires large investments, the world is predicted to be unavoidably increased in the prices for motor fuel. In terms of increased requirements for environmental protection from harmful emissions from exhaust gases of internal combustion engines, there was also a serious problem with ensuring the quality of

DOI: 10.1201/9781003177593-17

motor fuels. The process of sulfur content reducing in diesel fuels has led to a loss of a number of consumer properties. Therefore, in order to improve the lubricating properties of the environmental-friendly diesel fuels, it is necessary to add the anti-wearing additives in them.

Thus, today there is a multifaceted topical problem of ensuring the needs of the Ukrainian and European automobile transport in high-quality and environmental-friendly diesel fuel. One of the main directions for solving this problem is the usage of renewable energy sources from plant biomass (Panchuk et al. 2017). At the same time, the rapid growth of the production and consumption of the biodiesel fuel from vegetable oils of food use in many countries of the world led to a disturbance of the balance in the structure of agro-industrial production and began to generate the problems of the socio-ethical and ecological plan.

One of the promising further ways of biodiesel fuel developing, increasing of efficiency, and reducing the diesel fuel costs is the usage of the biomass of algae (Kryshtopa et al. 2018b), which as energy raw materials exceed other raw bioresources by its characteristics. However, the widespread adoption of biofuel from algae as an additive to motor fuel is hampered by the insufficient study currently being undertaken on the use of biofuels in the automotive engines that are made from these biomaterials. Therefore, research on the use of biofuels in automobile engines created from a large range of existing algae is opportune and relevant.

17.2 ANALYSIS OF RECENT RESEARCH AND PUBLICATIONS

The biodiesel fuel can be divided into two groups. The first one can include mixed fuels, which consist of petroleum diesel fuel with additives of non-petroleum origin fuels. Mixed diesel fuels are as a rule akin to traditional oil fuels according to performance indicators (Matijošius et al. 2009).

There are well-known works of diesel fuels mixtures researchers with liquefied hydrogen and Brown gas. However, the hydrogen power plant on the basis of the traditional diesel ICE is much more difficult and more expensive to service than conventional diesel ICE. According to the Massachusetts Institute of Technology, usage of a hydrogen car at this stage of hydrogen technology development is a hundred times more expensive than diesel (Juknelevičius et al. 2019). In addition, to refuel with hydrogen, it is necessary to build a network of filling stations.

For the gas stations refueling with hydrogen, the price of equipment is much higher than for the filling stations that refuel cars with liquid fuels (gasoline, ethanol and diesel fuel). According to GM data, the construction of hydrogen fuel stations is estimated at about $ 1 million per one gas station (Rimkus et al. 2013) while the equipment set for the conventional gas stations costs from $ 40,000, with an average of $ 100,000–200,000. It is also necessary to note rather a high price of such mixed fuel: about 8 €/l (Lebedevas et al. 2012) since at the moment the hydrogen is produced by consuming a significant amount of electricity.

A number of studies of usage of diesel fuel mixtures with gasoline, gas condensate, and turpentine have been carried out (Butkus et al. 2007). The disadvantage of these mixed fuels is forming of resins which coke and disable the engine during the process of their burning.

The second group includes synthetic fuels that differ from traditional petroleum ones due to their physical and chemical characteristics; therefore, an adaptation of technology is required for their using. The biodiesel fuels in particular belong to such kinds of fuels.

Under the conditions of modern production, biofuels are derived from vegetable oils by the reaction of transesterification (Awolu & Layokun 2013). This reaction does not require complicated process equipment and high temperatures, and the resulting mixture of esters is little different from hydrocarbons of petroleum fuels with the best environmental and lubricating properties (Wang et al. 2015).

Using the biodiesel from land crops, a significant reduction in emissions of particulate matter (soot), carbon monoxide (II), and hydrocarbons (including carcinogens) was found to be significant in comparison with the use of petroleum fuels (Eggert & Greaker 2014).

The high temperature of fuel ignition (above 120°C) makes the usage, storage, and transportation of the biodiesel fuel safer in comparison with the diesel fuel of petroleum origin (Ghosh 2015). Another significant advantage of biodiesel is its ability to biodegrade, unlike petroleum (Marinescu & Cicea 2018).

Various types of vegetable oils such as canola, flax, sunflower, palm, and others are used in the production of biodiesel fuel (German et al. 2011). In this case, biofuel from different vegetable oils has a number of distinct physical and chemical characteristics. Such signs include lower heat of combustion, viscosity, density, filtering, the temperature of freezing, coking, cetane number, etc. (Makarevičienė et al. 2013). Rape is the optimum raw material for the biodiesel production. The percentage of biodiesel fuel output from 1 ton of rapeseed oil is 96%. However, this fuel has a short shelf life, properties of the solvent; therefore, it is aggressive to engine details; at low temperatures, there is a precipitate, which leads to blockage of details and pollution of filters (Makarevičienė et al. 2013).

The fuel potential of oilseeds, when compared to 1 ton of raw material, is much greater than other crops. Assuming that energy consumption for rapeseeds production is 17,700 MJ/ha, for oil extraction is 700 MJ/ha and the amount of energy recovered from oil is 22,200 MJ/ha, it can be calculated that the energy gained per hectare of rapes sown is 3,800 MJ (corresponding to 110 l of petroleum diesel fuel at its energy value) (Scharff et al. 2013, Dragobetskii et al. 2015).

Biofuels from terrestrial crops (rape, sunflower, etc.) are successfully used in the existing engines, extending the life of engines, and having a high cetane number (Haas & Wagner 2011). In the work by Kryshtopa et al. (2017), the results of experimental research on the parameters of diesel engines working on a mixture of diesel fuel and fusel oils are resulted. The usage of biofuels as a bio additive to petroleum diesel can improve the environmental and anti-wear properties of fuels (Nascimento et al. 2013).

Biodiesel fuel is an alternative energy source that is suitable for all vehicles with conventional diesel engines (Lingaitis & Pukalskas 2008a, Makarevičienė, Matijošius et al. 2013). The cost of producing biodiesel fuel in many countries is higher than the cost of petroleum products. Nowadays, the development of the biofuel market in Ukraine from land crops in the absence of subsidies from the state is economically inefficient (Lingaitis & Pukalskas 2008b).

Biofuels slightly change the characteristics of power and consumption of diesel engines. The power of the diesel engine on biofuels drops by 7%–8%, and the fuel consumption increases by about 10% compared with conventional diesel fuel (Lingaitis & Pukalskas 2007).

A promising direction of reducing the diesel fuel cost is the conversion of the diesel engines to work on gas fuels (Kryshtopa et al. 2018a). The conversion of diesel engines to gaseous fuels can simultaneously reduce the toxicity of diesel engine exhaust in particular to reduce emissions of nitrogen oxides.

While using algae as biomaterials for the production of motor fuels, there are many advantages (Behera et al. 2015): during the growth process, algae absorb 80%–90% of carbon dioxide with oxygen release; for the cultivation of algae, one can use waste and saline water; algae, unlike terrestrial plants, grow year-round. It was established that the bioavailability and lipid content of algae both depend on the intensity of light (Afify et al. 2010). High intensity of light leads to the accumulation of lipids in algae. Algae consume mainly light in red and blue ranges.

Low mixing of water intensifies the heat and mass transfer processes in algae, promotes the movement of cells into the area of illumination, and increases the bio-productivity of algae (Chen et al. 2013). It was established that the concentration of carbon dioxide (Ho et al. 2011) has a significant effect on the yield of algae. Therefore, increasing the carbon concentration of dioxide from 4% to 22% enables to increase the biomass yield of algae from four to five times (Ogorodnikov et al. 2004, Ogorodnikov et al. 2018).

Algae are the oldest and most enduring organisms on the Earth, which were formed about 350 million years ago and live in fresh and salt water, soil and even snow. Among all the diversity of the existing algae, the authors of the study selected the Cyanophyta family Chroococcales blue-green algae. These algae lead to intensive "blooming" of reservoirs in Ukraine and Europe, with each year the extent of pollution of the surface of water significantly increases. The proliferation of blue-green algae leads to water rotting, the destruction of aquatic ecosystems and the destruction of rivers and lakes. The most effective way to clean the reservoirs is to use algae as fuel (Behera et al. 2015, Vorobyov et al. 2017). Therefore, the purposes of this chapter are experimental studies:

- changes in the performance characteristics of automotive diesel engines when used in these diesel fuel oil engines and their mixtures with biofuels derived from blue-green algae.
- changes in the environmental performance of automotive diesel engines with the use of petroleum diesel and their mixtures with biofuels derived from blue-green algae.

17.3 METHODS AND MATERIALS

The stand tests were carried out on an experimental installation, which included a D21A1 series diesel, a short technical characteristic of which is given in Table 17.1. The scheme and appearance of the D21A1 diesel engine are shown in Figure 17.1.

The loading for the D21A1 engine (5) is created by a four-stage four-cylinder compressor of the brand K-5M (9). The power of the K-5M compressor shaft can be adjusted in the range of 1–35 kW, which allows the diesel engine to be 100% loaded. The torque from the D21A1 engine is transmitted to the compressor using the gearbox (5) and the gearbox (7).

Table 17.1 Brief technical characteristics of the D21A1 experimental diesel engine

Name of engine parameters	*Unit of measurement*	*Value*
Type of diesel	–	Four-stroke, two-cylinder, air-cooled
Working volume	l	2.08
Method of mixing	–	Separate combustion chamber with direct injection of diesel fuel
Nominal engine power	kW (hp)	18 (25)
Effective specific fuel consumption	g/kW · h (g/hp · h)	253 (186)
Frequency of the crankshaft rotation of engine at nominal power	rpm	1,800
Frequency of the crankshaft rotation of engine at idle speeds	rpm	800
The mass of diesel engine	kg	280

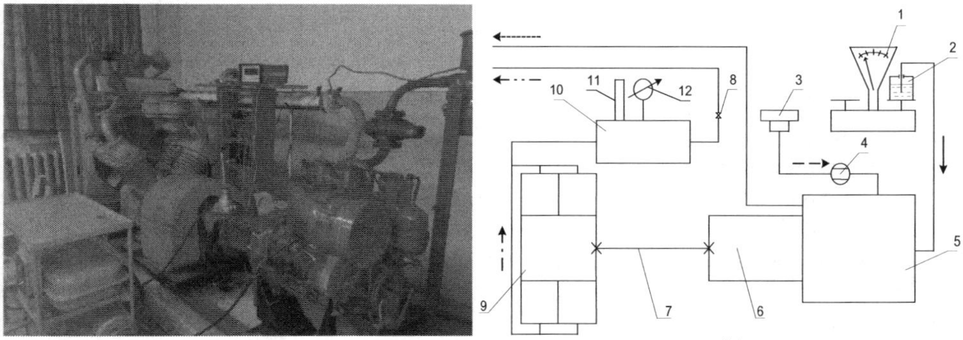

(a) (b)

Figure 17.1 Appearance (a) and scheme of the experimental installation on the basis of the D21A1 diesel engine (b) for studding of engine indicators on biodiesel mixtures – the direction of fuel movement in the power system of experimental installation; – direction of air movement in the power system of experimental installation; – airflow to the receiver; – the motion of exhaust gases into the environment; – the motion of air into the environment; 1 – scales for measuring of fuel consumption; 2 – volume for biodiesel fuel; 3 – air filter; 4 – gasometer; 5 – experimental diesel engine; 6 – gearbox; 7 – Cardan transmission; 8 – throttle; 9 – compressor; 10 – receiver; 11 – the air temperature thermometer; and 12 – manometer.

The effective power N_e, kW was determined by the formula

$$N_e = \pi \cdot \frac{M_e \cdot n_x}{3 \times 10^4 \cdot i}, \tag{17.1}$$

where

M_e is the effective wheel torque, Nm;
n_x is the engine speed, $\min^{-1}$;
i is the transmission ratio.

The indicated specific fuel consumption g, g/kWh was experimentally determined through the engine mechanical efficiency η_{en} and transmission η_{tr}, hourly fuel consumption G and effective wheel power N_e as follows:

$$g = G/(N_e \cdot \eta_{en} \cdot \eta_{tr}). \tag{17.2}$$

The engine mechanical efficiency η_{en} and transmission η_{tr} were determined through the mechanical loss power of the engine N_{en} and the mechanical loss power of the transmission N_{tr}

$$\eta_{en} = \frac{N_e}{N_e + N_{en}}, \tag{17.3}$$

$$\eta_{tr} = 1 + \frac{N_e + N_{en}}{N_{tr}}. \tag{17.4}$$

The mechanical loss powers were determined when measuring the power consumption of the electric motor, which was interlocked by the retainer with the drive wheel. The mechanical loss power of the transmission N_{tr} was measured in the neutral position of the gearbox. The mechanical loss power of the engine N_{en} and the mechanical loss power of the transmission N_{tr} were measured in the highest gear.

Blue-green algae were collected from the reservoirs around the IvanoFrankivsk region (Figure 17.2a) for harvesting biofuels in the summertime, and in winter, they were cultivated in phyto-bioreactors (Figure 17.2b).

Photo-bioreactor is a 60 l transparent tank illuminated with fluorescent lamps. Algae cultivate with cycles of 14 days. The gasair mixture inputs into this reactor using a compressor and a carbonaceous cylinder on a volume of 1 cubic meter in a day (carbon dioxide – 8%). Drying of biomass was carried out in a drying cabinet to a humidity level of 10%.

The additive was obtained by the reaction of methanolysis of vegetable oils in the presence of a homogeneous catalyst. Potassium hydroxide (4.3%) solution in methyl alcohol (95.7%) was used for its preparation. Then, the prepared solution and the lipid components of the algae were fed into the apparatus, where the mixing of the reaction products and the synthesis of the supplement took place. Further, the reaction mass was separated in a separator from glycerol. The resulting bio-additives were purified with an aqueous solution of orthophosphoric acid, and the residues of water evaporated. The resulting additive was mixed with petroleum diesel fuel.

Calculations showed that the combustion heat of obtained methyl esters is 35.5 MJ/kg. It should be noted that after extraction of fats, dry biomass can be additionally used with a combustion heat of 15.5 MJ/kg.

(a) (b)

Figure 17.2 Sources of harvesting of blue-green algae for biofuel production: natural reservoirs (a) and artificial bioreactors (b).

Thus, in experimental studies, the diesel fuel mixtures with derived bioactive compounds based on methyl esters of Chroococcales, blue-green algae, in quantities of 5%, 10%, and 20% and for comparison pure diesel fuel were used. In this case, the fuel tank was filled with diesel fuel of mark L of the Kremenchug oil refinery.

The volumetric particles of carbon monoxide and hydrocarbons were measured by the "Autotest-02.03P" gas analyzer. The range of measurements of hydrocarbon gas analyzer is 0–2,000 ppm, the absolute measurement error is 10 ppm. The range of measurement of carbon monoxide gas analyzer is 0%–5%, and the absolute measurement error is 0.03%. The thermocouples of the type "chromel-copel" and the log-meter-potentiometer UP-2M were used for finding the temperatures of exhaust gases.

For comparative estimation of the engine indices on diesel fuel with corresponding indicators of a diesel engine on a mixture of diesel fuel with bioactive additives on the basis of methyl esters of the lipid fraction of blue-green algae in the amount of 5%, 10%, and 20%, the load characteristics of the engine were removed at fixed speeds of the crankshaft. Before measuring the parameters for stable running of the working process, the engine worked at least 5 minutes in the given mode. The results of the measurements were entered in the protocol of trials with a threefold repetition at each mode of the diesel engine operation.

17.4 RESULTS OF THE EXPERIMENTAL RESEARCH

As a result of the performed experimental research, the dependences characterizing the change of the effective engine power using diesel fuel and a mixture of diesel fuel with the received bioactive supplements based on methyl esters of the lipid fraction of the Chroococcales blue-green algae in the amount of 5%, 10%, and 20% (Figure 17.3) were determined. It was experimentally established that the effective power of an engine using a mixture of diesel fuel with the derived bioactive compounds based on methyl esters of the lipid fraction of Chroococcales blue-green algae in the amount of 5%, 10%, and 20% will decrease by an average of 0.9%, 1.8%, and 3.5%.

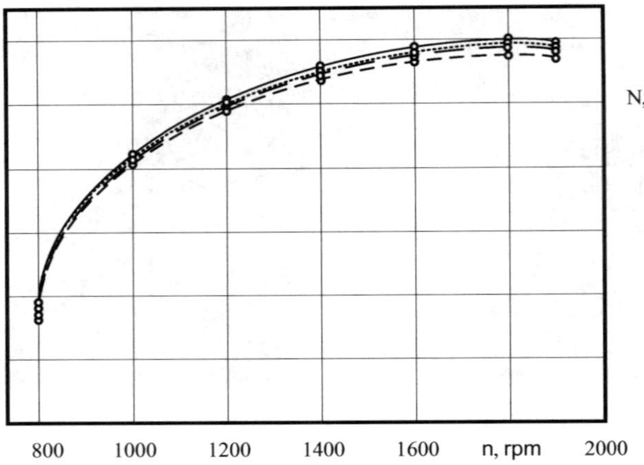

Figure 17.3 Experimental dependencies of effective power *N* with respect to crankshaft rotational frequency n of the various contents of bioadditives: ——— engine work on 100% petroleum diesel fuel; engine work on the mixture of 95% petroleum diesel and 5% methyl esters; – – – – engine work on a mixture of 90% of petroleum diesel and 10% of methyl esters; - - - - - engine work on a mixture of 80% of petroleum diesel and 20% of methyl esters.

As a result of the performed experimental research the following relations were obtained:

- dependence of changes in the CnHm hydrocarbons contents on the crankshaft speed of the engine n at the nominal loading for the various content contents of the supplements on the basis of methyl esters of the lipophilic fraction of the Chroococcales blue-green algae in the amount of 5%, 10%, and 20% (Figure. 17.4);
- dependence of the change in the content of carbon monoxide CO on the rotational speed of the crankshaft engine n at the nominal load for the various contents of bioadditives on the basis of methyl esters of the lipid fraction of the Chroococcales blue-green algae in the amount of 5%, 10%, and 20% (Figure. 17.5).

It was experimentally established that the content of C_nH_m hydrocarbons using a mixture of diesel fuel with the derived supplements based on methyl esters of the lipid fraction of the Chroococcales blue-green algae in the amount of 5%, 10%, and 20% will decrease by an average of 6.2%, 13.1%, and 26.6%. It was experimentally determined that the content of carbon monoxide in the use of a mixture of diesel fuel with the derived bioactive compounds based on methyl esters of the lipid fraction of the Chroococcales blue-green algae in the amount of 5%, 10%, and 20% will decrease by an average of 6.5%, 13.9%, and 28.7%.

According to the results of the experiments, the value of the indicated specific fuel consumption on 100% petroleum diesel fuel at the nominal mode was 220 g/kWh, and the minimum value of the indicated specific fuel consumption was 185 g/kWh. According to the results of the experiments, the value of the indicated specific fuel consumption on a mixture of 80% of petroleum diesel and 20% of methyl esters at the

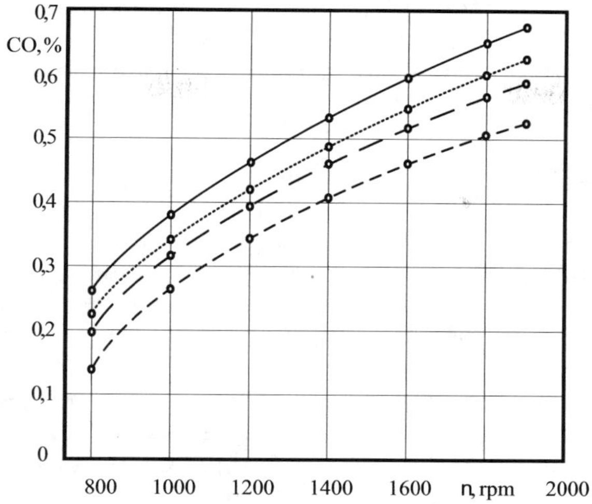

Figure 17.4 Experimental dependences of the hydrocarbons C_nH_m content with respect to the rotational frequency of the crankshaft engine n at nominal loading for different contents of bioadditives: ——— engine work on 100% petroleum diesel fuel; ········ engine work on a mixture of 95% petroleum diesel and 5% methyl esters; – – – engine work on a mixture of 90% of petroleum diesel and 10% of methyl esters; and - - - engine work on a mixture of 80% of petroleum diesel and 20% of methyl esters.

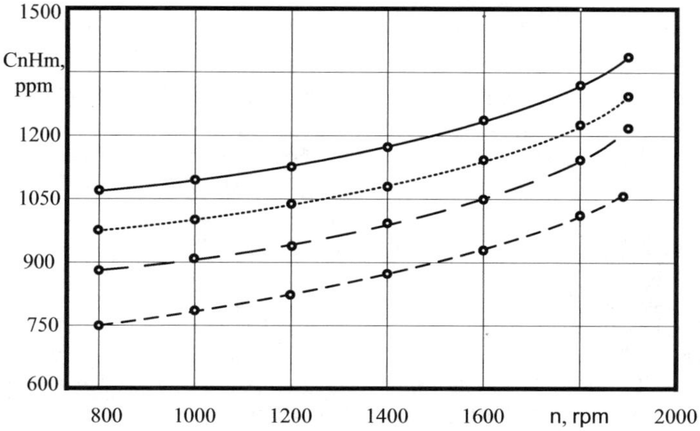

Figure 17.5 Experimental dependencies of the carbon monoxide CO content with respect to crankshaft rotational frequency *n* at nominal loads for different contents of bioadditives: ——— engine work on 100% petroleum diesel fuel; ········ engine work on a mixture of 95% petroleum diesel and 5% methyl esters; – – – engine work on a mixture of 90% of petroleum diesel and 10% of methyl esters; and - - - engine work on a mixture of 80% of petroleum diesel and 20% of methyl esters.

nominal mode was 238 g/kWh, and the minimum value of the indicated specific fuel consumption was 203 g/kWh. Thus, on a mixture of 80% of petroleum diesel and 20% of methyl esters, the experimental value of the indicated specific fuel consumption, depending on the changes in the crankshaft speed increased in comparison with the first option in the range from 8.1% to 9.7%.

17.5 CONCLUSIONS

Synthetic methyl ester of lipid fraction was used as an additive of 5%–20% to conventional diesel fuel. In the course of experimental tests, it was established that on different blends of biodiesel and diesel fuels, the difference in power characteristics is observed in the range of 0.9%–3.5%, which from the operational point of view is not a significant difference.

At the same time, increasing of bio-additives in the biodiesel fuel environmental significantly improved the performance of engines and the contents of unburnt hydrocarbons and carbon monoxides was significantly reduced. The lowest rates were recorded when using a biodiesel mixture containing 20% of bio-additives. The contents of hydrocarbons and oxides of carbon in diesel fuel with bio-additives on the basis of methyl esters of the lipid fraction of the Chroococcales blue-green algae decreased by 26.6% and 28.7%, respectively.

The obtained results allow optimizing the choice of fuels for power systems of internal combustion engines and reducing the emissions of harmful substances in the exhaust gases of automobile diesel engines. Further research will be related to the definition of fuel consumption and emissions of nitrogen oxides in the exhaust gases of diesel engines converted to work on alternative fuels with bioactive additives based on methyl esters of algae.

REFERENCES

Afify, A.M.M., Shanab, S.M. & Shalaby, E.A. 2010. Enhancement of biodiesel production from different species of algae. *Grasas y Aceites* 61: 416–422.

Awolu, O.O. & Layokun, S.K. 2013. Optimization of two-step transesterification production of biodiesel from neem (Azadirachta indica) oil. *International Journal of Energy and Environmental Engineering.* 4(1): 1–9.

Behera, S., Singh, R., Arora, R., Sharma, N., Shukla, M. & Kumar, S. 2015. Scope of algae as third generation biofuels, *Frontiers in Bioengineering and Biotechnology* 2: 90.

Butkus, A., Pukalskas, S. & Bogdanovičius, Z. 2007. The influence of turpentine additive on the ecological parameters of diesel engines. *Transport* 22(2): 80–82.

Chen, C.Y., Zhao, X.Q., Yen, H.W., Ho, S.H., Cheng, C.L. & Bai, F. 2013. Microalgae-based carbohydrates for biofuel production. *Biochemical Engineering Journal* 78: 1–10.

Dragobetskii V., Shapoval A., Mos'pan D., Trotsko O. & Lotous V. 2015. Excavator bucket teeth strengthening using a plastic explosive deformation. *Metallurgical and Mining Industry* 4: 363–368.

Eggert, H. & Greaker M. 2014. Promoting second generation biofuels: Does the first generation pave the road?. *Energies* 7(7): 4430–4445.

German, L., Schoneveld, G.C. & Pacheco, P. 2011. The social and environmental impacts of biofuel feedstock cultivation: Evidence from multi-site research in the forest frontier. *Ecology and Society* 16(3): 24.

Ghosh, P. 2015. Biofuels: Engineering and biological challenges. *Proceedings of Indian National Science Academy* 81(4): 765–773.

Haas, M.I. & Wagner, K. 2011. Simplifying biodiesel production: The direct or in situ transesterification of algal biomass. *European Journal of Lipid Science and Technology* 113: 1219–1229.

Ho, S.H., Chen, C.Y., Lee, D.J. & Chang, J.S. 2011. Perspectives on microalgal CO_2-emission mitigation systems – A review. *Biotechnology Advances* 29: 189–198.

Juknelevičius, R., Rimkus, A., Pukalskas, S. & Matijošius, J. 2019. Research of performance and emission indicators of the compression-ignition engine powered by hydrogen – Diesel mixtures. *International Journal of Hydrogen Energy* 44(20): 10129–10138.

Kryshtopa, S., Kryshtopa, L., Melnyk, V., Prunko, I., Dolishnii, B. & Demianchuk, Y. 2017. Experimental research on diesel engine working on a mixture of diesel fuel and fusel oils. *Transport Problems* 12(2): 53–63.

Kryshtopa, S., Panchuk, M., Dolishnii, B., Kryshtopa, L., Hnyp, M. & Skalatska, O. 2018a. Research into emissions of nitrogen oxides when converting the diesel engines to alternative fuels. *Eastern-European Journal of Enterprise Technologies* 10(91): 16–22.

Kryshtopa, S., Panchuk, M., Kozak, F., Dolishnii, B., Mykytii, I. & Skalatska, O. 2018b. Fuel Economy raising of alternative fuel converted diesel engines. *Eastern-European Journal of Enterprise Technologies* 8(94): 6–13.

Lebedevas, S., Pukalskas, S., Žaglinskis, J. & Matijošius, J. 2012. Comparative investigations into energetic and ecological parameters of camelina-based biofuel used in the diesel engine. *Transport* 27(2): 171–177.

Lingaitis, L.P. & Pukalskas, S. 2007. Determining the consumption of biodiesel by locomotive engines. *Transport Means 2007: Proceedings of the 11th International Conference*: 194–197, Kaunas, Lithuania.

Lingaitis, L.P. & Pukalskas, S. 2008a. Ecological aspects of using biological diesel oil in railway transport. *Transport* 23(2): 138–143.

Lingaitis, L.P. & Pukalskas, S. 2008b. The economic effect of using biological diesel oil on railway transport. *Transport* 23(4): 287–290.

Makarevičienė, V., Matijošius, J., Pukalskas, S., Vėgneris, R., Kazanceva, I. & Kazancev, K. 2013. The exploitation and environmental characteristics of diesel fuel containing rapeseed butyl esters. *Transport* 28(2): 158–165.

Makarevičienė, V., Sendzikiene, E., Pukalskas, S., Rimkus, A. & Vėgneris, R. 2013. Performance and emission characteristics of biogas used in diesel engine operation. *Energy Conversion and Management* 75: 224–233.

Marinescu, C. & Cicea, C. 2018. What do we know about biofuels today that Diesel and Ford did not. *Management and Economics Review* 3(2): 213–224.

Matijošius, J., Mažeika, M. & Pukalskas, S. 2009. Calculation methodology of working cycle parameters of the diesel engine operating on multicomponent mixture. *Rural Development 2009: Proceedings of the Fourth International Scientific Conference*, 15–17 October, 2009, Akademija, Kaunas Region, 4(2): 355–359, Akademija, Lithuanian University of Agriculture, Lithuania.

Nascimento, I.A., Marques, S.S.I., Cabanelas, I.T.D., Pereira, S.A., Druzian, J.I. & de Souza, C.O. 2013. Screening microalgae strains for biodiesel production: Lipid productivity and estimation of fuel quality based on fatty acids profiles as selective criteria. *BioEnergy Research* 6: 1–13.

Ogorodnikov, V.A., Dereven'ko, I.A. & Sivak, R.I. 2018. On the influence of curvature of the trajectories of deformation of a volume of the material by pressing on its plasticity under the conditions of complex loading. *Materials Science* 54(3): 326–332.

Ogorodnikov, V.A., Savchinskij, I.G. & Nakhajchuk, O.V. 2004. Stressed-strained state during forming the internal slot section by mandrel reduction. *Tyazheloe Mashinostroenie* 12: 31–33.

Panchuk, M., Kryshtopa, S., Shlapak, L., Kryshtopa, L., Yarovyi, V. & Sladkovskyi, A. 2017. Main trend of biofuels production in Ukraine. *Transport Problems* 12(4): 95–103.

Rimkus, A., Pukalskas, S., Matijošius, J. & Sokolovskij, E. 2013. Betterment of ecological parameters of a diesel engine using Brown's gas. *Journal of Environmental Engineering and Landscape Management* 21(2): 133–140.

Scharff, Y., Asteris, D., Fédou, S. 2013. Catalyst technology for biofuel production: Conversion of renewable lipids into biojet and biodiesel. *Oilseeds and Fats, Crops and Lipids* 20(5): 502.

Vorobyov V., Pomazan M., Vorobyova L. & Shlyk S. 2017. Simulation of dynamic fracture of the borehole bottom taking into consideration stress concentrator. *Eastern-European Journal of Enterprise Technologies* 3/1(87): 53 – 62.

Wang, M., Wu, W., Wang, S., Shi, X., Wu, F. & Wang. J. 2015. Preparation and Characterization of a solid acid catalyst from macro fungi residue for methyl palmitate production. *BioResources* 10(3): 5691–5708.

Chapter 18

Mathematical Model of Synthesis of Biodiesel from Technical Animal Fats

Mikhailo M. Mushtruk and Igor P. Palamarchuk
National University Life and Environmental Sciences of Ukraine

Vadim P. Miskov and Yaroslav V. Ivanchuk
Vinnytsia National Technical University

Paweł Komada
Lublin University of Technology

Maksat Kalimoldayev
Institute of Information and Computational Technologies CS MES RK

Karlygash Nurseitova
East Kazakhstan State Technical University named after D. Serikbayev

CONTENTS

18.1 INTRODUCTION

The high price of biodiesel (over double the price of diesel) mostly results from the price of the feedstock, which makes up to 75% of the overall production costs (Adewale, Dumont, J. & Ngadi 2015). However, biodiesel can be made from other feed stocks, including beef tallow, pork lard, yellow grease, and waste tannery fats (Chai et al. 2014, Farooq, Ramli, & Subbarao 2013, Hawrot-Paw, Wijatkowski, & Mikiciuk 2015), which are unpleasant wastes and are consequently incinerated. The incineration

DOI: 10.1201/9781003177593-18

is charged, which makes the tannery waste economically profitable in the biodiesel production. The acid value of flashings (the main waste fat produced by tanneries) usually exceeds 2 mg KOH/g.

Therefore, their direct processing via alkali-catalyzed transesterification is not suitable since the inorganic alkali catalyst (KOH, NaOH) is spent on free fatty acids (FFA) neutralization. Moreover, the salts formed during this reaction prevent easy separation of the methyl ester and glycerin phase; also, simultaneously formed water decreases the transesterification yield (Issariyakul & Dalai 2014, Mohammadshirazi et al. 2014). There are several methods capable to overcome this problem. Probably, the most investigated way is an acid catalyzed transesterification of FFA used at an industrial scale (Mushtruk et al. 2017), in contrast to the majority of other techniques (e.g. enzymatic transesterification and transesterification (Nielsen et al. 2016), fat treatment in supercritical methanol (Sajith, Sobhan, & Peterson 2010)).

The acid-catalyzed transesterification method uses a strong inorganic acid (like sulfuric acid) for transesterification of the FFA to methyl esters. The transesterification reduces the FFA content to a level at which an alkali catalyst may be employed for subsequent transesterification of glycerols (Sukhenko et al. 2017). In Sukhenko et al. (2018), a scale-up of the two-step transesterification technology is used, in which sulfuric acid was employed as a transesterification catalyst and the refined fat was transesterification with an inorganic alkali.

The main drawback of the described techniques is the necessity of acid neutralization prior to alkali transesterification, which leads to a formation of large amounts of salts in the reaction system (Xiong, Guo, & Xie 2015). These salts must be removed from the final biodiesel (and also glycerin), which may be the limiting factor during the product purification (Kolyanovska et al. 2019).

Thus, much effort is made to develop suitable solid transesterification catalysts (e.g. Xiong, Guo, & Xie 2015), which are easily separable from the reaction mixture after successful transesterification. Despite the fact that some of these solid catalysts are available commercially, the solid catalyst efficiency and its possible fouling are still an issue, namely in the case of transesterification under mild reaction conditions, that is, at atmospheric pressure and temperatures around 60°C (Xiong, Guo, & Xie 2015).

Modeling has become one of the most powerful means of making engineering decisions in modern technical re-equipment of production. With its application and introduction, there are undeniable advantages in reducing the terms of experimental research and saving resources in science and technology. Mathematical modeling of processes in the overwhelming majority of cases may rule out the need for pilot installations to test technical and technological solutions, or to carry out labor-intensive and expensive experiments (Adewale, Dumont, & Ngadi 2015).

There are many ways of performing mathematical modeling of technological processes. Its most important purpose is determined by the way of establishing the possibility of simulating the process with the help of physical, mathematical, chemical, and other models, separately or in combination with the laws, mechanisms, and connections between the influencing factors, in order to calculate the parameters of the model as accurately as possible (Chai et al. 2014).

The aim and objectives of research are to construct a mathematical model of the process of transesterification of technical animal fats in biodiesel.

18.2 LITERATURE REVIEW AND PROBLEM STATEMENT

The basics of the theory and results of experimental studies of technological processes, machines, and equipment for the production and use of diesel and other types of biodiesel in the agro industrial complex of Ukraine are presented in detail in the scientific works of M. Virevky, V. Voytova, Ya. Gukova, V. Dubrovina, B. Kochirky, S. Kovalishina, V. Kravchuka, M. Linnaeus, V. Mironenka, S. Pastushenko, G. Ratushniaka, V. Semenova, Yu. Sukhenka, G. Topilina, and V. Yasenetsky et al. (Issariyakul & Dalai 2014, Mohammadshirazi et al. 2014, Mushtruk et al. 2017).

However, little attention is paid to the study of the processes of mathematical modeling for the production of biodiesel, which reduces the cost and increases the yield and quality of the final product (biodiesel), although the volume of raw materials is rather high. Therefore, the relevance of the chosen direction of research is beyond doubt.

18.3 MATERIALS AND METHODS OF RESEARCH

Technical animal fats – triglycerides (TG) – consist of three chains of fatty acids, which are joined by molecules of glycerol.

When passing the reaction of transesterification with methyl (ethyl) alcohol, triglycerides are converted to methyl (ethyl) esters of fatty acids (biodiesel). Catalysts are used for intensification of the process (often sodium hydroxide or potassium). Hydroxides break the bonds in the triglyceride molecules and provide their reaction with methanol or ethanol to form esters that are biodiesel. Other catalysts may also be used (Farooq, Ramli, & Subbarao 2013, Issariyakul & Dalai 2014, Sukhenko et al. 2017).

At the end of the re-transesterification, the triglyceride molecules are converted into three molecules of methyl ether and one molecule of glycerol.

This reaction takes place in three stages (Figure 18.1). First, a chain of fatty acids is separated from the triglyceride (TG) and is joined to methanol (A), forming a molecule of methyl ether (ME) and diglyceride (DG). Afterwards, the second chain of fatty acid is separated from diglycerides and binds to the methanol molecule to form a molecule of methyl ether and monoglycerides (MG). Then, in monoglycerides, glycerol is replaced with methyl alcohol to form methyl ester and glycerol. This step completes the reaction (Hawrot-Paw, Wijatkowski, & Mikiciuk 2015).

Initially, the reaction proceeds very quickly, but over time, the pace of conversion slows down. Some triglycerides become diglycerides with the release of methyl esters of fatty acids (biodiesel). Diglycerides are more slowly converted into monoglycerides, which are then more slowly converted into methyl esters, and there is a complete loss of free glycerin in the precipitate. The rate of reaction decreases gradually, and it never ends completely. At the end of the process, the amount of glycosides is negligible and limited to the strict limits set by the standard for biodiesel fuels (Issariyakul & Dalai 2014).

Each of the stages of the transesterification reaction can take place in the forward and reverse direction with an individual rate, which depends on the reaction rate constant k_n.

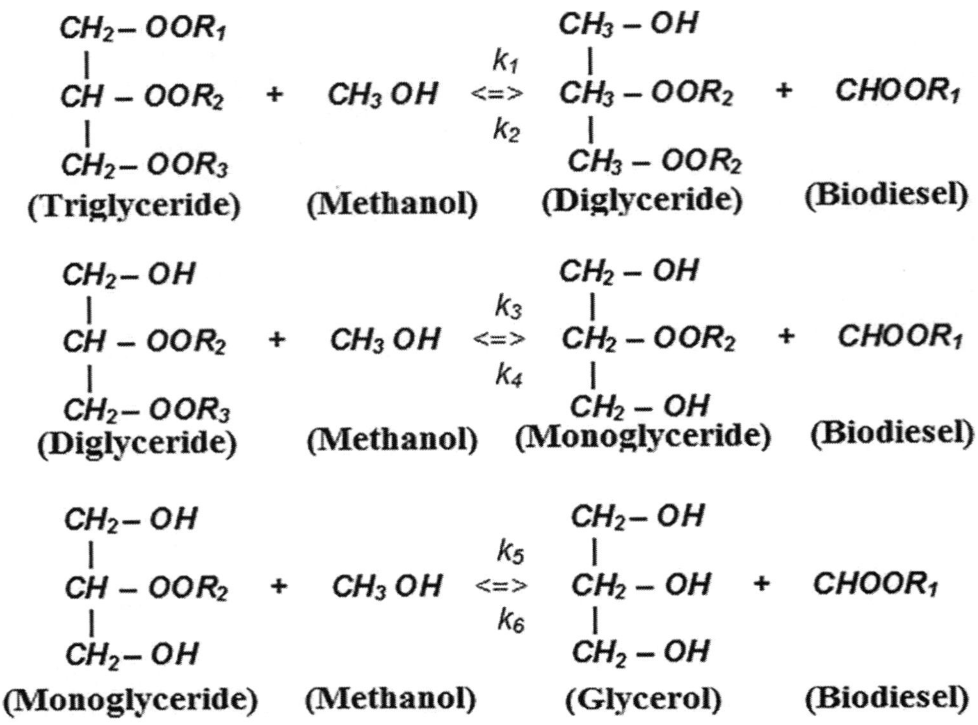

Figure 18.1 Scheme of phased transesterification of triglycerides.

In order to determine the rate of occurrence and disappearance of the components of the reacting mixture, the following system of differential equations that collectively simulates the process of transesterification of triglycerides of fat can be used:

$$\frac{d[\mathrm{TG}]}{dt} = -k_1[\mathrm{TG}] + k_2[\mathrm{DG}][\mathrm{E}] \tag{18.1}$$

$$\frac{d[\mathrm{DG}]}{dt} = -k_3[\mathrm{DG}][\mathrm{A}] + k_4[\mathrm{MG}][\mathrm{E}] + k_1[\mathrm{TG}][\mathrm{A}] - k_2[\mathrm{DG}][\mathrm{E}]; \tag{18.2}$$

$$\frac{d[\mathrm{MG}]}{dt} = -k_5[\mathrm{GL}][\mathrm{E}] + k_6[\mathrm{GL}][\mathrm{E}] + k_3[\mathrm{DG}][\mathrm{A}] - k_4[\mathrm{MG}][\mathrm{E}]; \tag{18.3}$$

$$\frac{d[\mathrm{GL}]}{dt} = k_5[\mathrm{MG}][\mathrm{A}] - k_6[\mathrm{GL}][\mathrm{E}]; \tag{18.4}$$

$$\frac{d[\mathrm{E}]}{dt} = k_1[\mathrm{TG}][\mathrm{A}] - k_2[\mathrm{DG}][\mathrm{E}] + k_3[\mathrm{DG}][\mathrm{A}] - k_4[\mathrm{MG}][\mathrm{E}] + k_5[\mathrm{MG}][\mathrm{A}] - k_6[\mathrm{GL}][\mathrm{E}]; \tag{18.5}$$

$$\frac{d[A]}{dt} = -k_1[TG][A] - k_2[DG][E] + k_3[DG][A] - k_4[MG][E] + k_5[MG][A] - k_6[GL][E]; \quad (18.6)$$

$$\frac{d[A]}{dt} = -\frac{d[E]}{dt}, \quad (18.7)$$

where the expressions in brackets indicate the molar concentrations of the following components:

[TG] – triglycerides,
[DG] – diglycerides,
[MG] – monoglycerides,
[GL] – glycerol,
[A] – alcohol,
[E] – methyl ether,
$k_1, k_2, k_3, k_4, k_5, k_6$ – reaction constants in the directions indicated in Figure 18.1.

To solve this model system of differential equations, a numerical method using the Microsoft Excel software was used. The values of the reaction constants are taken from the literature (Hawrot-Paw, Wijatkowski, & Mikiciuk 2015) and are given in Table 18.1.

At a fixed temperature, the reaction is possible if the interacting molecules have a certain amount of excess energy. Arrhenius called this excess energy the activation energy, and the molecules activated themselves. By Arrhenius, the constant of the reaction rate k and the activation energy E_a are related by the Arrhenius equation (Mohammadshirazi et al. 2014, Polishchuk et al. 2018, Polishchuk et al. 2019):

$$k = A \cdot e^{-Ea/RT}, \quad (18.8)$$

where,
E_a – value of the activation energy,
A – pre-exponential coefficient,
R – universal gas constant (8.31 J/mol·K),
T – absolute temperature, K.

Thus, at constant temperature, the reaction rate determines the level of activation energy E_a. The more E_a, the lower the number of active molecules and the slower the reaction proceeds. With decreasing E_a, the velocity increases, and at $E_a = 0$, the reaction proceeds instantaneously.

Table 18.1 Concentrations of reactions at a temperature of 50°C (mol% min)$^{-1}$ for animal fat

Constants	k_1	k_2	k_3	k_4	k_5	k_6
Size	0.05	0.11	0.215	1.228	0.242	0.007

Table 18.2 Activation energy, calcium $(\text{mole}\cdot\text{K})^{-1}$

Activation energy	E_1	E_2	E_3	E_4	E_5	E_6
Size	13,145	9,932	19,860	14,639	6,421	9,588

The quantity E_a characterizes the nature of the reactants and is determined experimentally with the dependence $k=f(T)$. By writing equation (18.8) in logarithmic form (18.9) and solving it for constants at two temperatures (18.10), we find E_a (Hawrot-Paw, Wijatkowski, & Mikiciuk 2015).

$$\ln\frac{k_1}{k_2}=\frac{(T_2-T_1)}{RT_2\cdot T_1}E_a \tag{18.9}$$

$$Ea=\frac{RT_1T_2\ln\left(\frac{kT_2}{kT_1}\right)}{T_1-T_2} \tag{18.10}$$

The results of the calculations of the activation energy in the reactions corresponding to the constants (see Table 18.1) are given in Table 18.2.

The output system of differential equations (18.1–18.7) can be transformed into the following system of discrete equations:

$$\frac{\text{TG}^{P+1}-\text{TG}^P}{\Delta t}=-k_1\text{TG}^P\text{A}^P+k_2\text{DG}^P\text{E}^P; \tag{18.11}$$

$$\frac{\text{DG}^{P+1}-\text{DG}^P}{\Delta t}=k_1\text{TG}^P\text{A}^P-k_2\text{DG}^P\text{E}^P-k_3\text{DG}^P\text{A}^P+k_4\text{MG}^P\text{E}^P; \tag{18.12}$$

$$\frac{\text{MG}^{P+1}-\text{MG}^P}{\Delta t}=k_3\text{DG}^P\text{A}^P-k_4\text{MG}^P\text{E}^P-k_3\text{MG}^P\text{A}^P+k_6\text{GL}^P\text{E}^P; \tag{18.13}$$

$$\frac{\text{GL}^{P+1}-\text{GL}^P}{\Delta t}=k_5\text{MG}^P\text{A}^P+k_6\text{GL}^P\text{E}^P; \tag{18.14}$$

$$\frac{\text{E}^{P+1}-\text{E}^P}{\Delta t}=-k_1\text{TG}^P\text{A}^P-k_2\text{DG}^P\text{E}^P+k_3\text{DG}^P\text{A}^P-k_4\text{MG}^P\text{E}^P+k_5\text{MG}^P\text{A}^P-k_6\text{GL}^P\text{E}^P; \tag{18.15}$$

$$\frac{\text{A}^{P+1}-\text{A}^P}{\Delta t}=-k_1\text{TG}^P\text{A}^P+k_2\text{DG}^P\text{E}^P-k_3\text{DG}^P\text{A}^P+k_4\text{MG}^P\text{E}^P-k_5\text{MG}^P\text{A}^P+k_6\text{GL}^P\text{E}^P, \tag{18.16}$$

where

p – the interval (dimensionless),
E – the concentration of methyl ether, mole,
Δt – step by time, seconds.

The current time t is determined by the formula:

$$t = p \cdot \Delta t; \tag{18.17}$$

The system of discrete equations (18.11–18.16) can be rewritten as follows:

$$\mathrm{TG}^{P+1} = \left(1 - k_1 \Delta t \mathrm{A}^P\right)\mathrm{TG}^P + k_2 \Delta t \mathrm{DG}^P \mathrm{E}^P; \tag{18.18}$$

$$\mathrm{DG}^{P+1} = \left(1 - k_2 \Delta t \mathrm{E}^P - k_3 \Delta t \mathrm{A}^P\right)\mathrm{DG}^P + k_1 \Delta t \mathrm{TG}^P \mathrm{A}^P + k_4 \Delta t \mathrm{MG}^P \mathrm{E}^P; \tag{18.19}$$

$$\mathrm{MG}^{P+1} = \left(1 - k_4 \Delta t \mathrm{E}^P - k_3 \Delta t \mathrm{A}^P\right)\mathrm{MG}^P + k_1 \Delta t \mathrm{DG}^P \mathrm{A}^P + k_6 \Delta t \mathrm{GL}^P \mathrm{E}^P; \tag{18.20}$$

$$\mathrm{GL}^{P+1} = \left(1 - k_6 \Delta t \mathrm{E}^P\right)\mathrm{GL}^P + k_5 \Delta t \mathrm{MG}^P \mathrm{A}^P; \tag{18.21}$$

$$\begin{aligned}\mathrm{E}^{P+1} = &\left(1 - k_2 \Delta t \mathrm{DG}^P - k_4 \Delta t \mathrm{MG}^P - k_6 \Delta t \mathrm{GL}^P\right)\mathrm{E}^P \\ &+ \left(k_1 \Delta t \mathrm{TG}^P + k_3 \Delta t \mathrm{DG}^P + k_5 \Delta t \mathrm{ML}^P\right)\mathrm{A}^P;\end{aligned} \tag{18.22}$$

$$\begin{aligned}\mathrm{A}^{P+1} = &\left(1 - k_1 \Delta t \mathrm{TG}^P - k_3 \Delta t \mathrm{DG}^P - k_5 \Delta t \mathrm{ML}^P\right)\mathrm{A}^P \\ &+ \left(k_2 \Delta t \mathrm{DG}^P + k_4 \Delta t \mathrm{MG}^P + k_6 \Delta t \mathrm{GL}^P\right)\mathrm{E}^P;\end{aligned} \tag{18.23}$$

This system of discrete equations has a solution provided that all the coefficients of the equations are equal to or less than one.

18.4 RESEARCH RESULTS

In order to calculate the developed model, the initial concentration of the reacting mixture in the ratio of 6 moles of methanol per mole of triglycerides was taken. The results of the calculations performed using the Microsoft Excel spreadsheet show (Table 18.3) that the reaction is fast during the first 2,500 seconds (0.69 hours), and then the rate of biofuel production is significantly slowed down, which correlates with the data of work (Mushtruk et al. 2017, Kozlov et al. 2019).

Figure 18.2a and b shows the effect of temperature on the yield of methyl ether. Obviously, an increase in temperature above 60°C practically does not affect the output of biodiesel.

The experimental studies on the degree of transformation of animal fats into biodiesel in time and their comparison with the data obtained theoretically using the mathematical model allowed confirming that the theoretical results fall into the confidence interval with the probability of 0.95 (Ogorodnikov, Zyska, & Sundetov 2018).

The analysis of the discrete system of equations generated by equations 18.18–18.23 enables to predict (see Figure 18.2a and b) an increase in the yield of

Table 18.3 Change in the concentration of components in the transesterification reaction over time at a temperature of 65°C

Reaction time, seconds	*Components of the reacting mixture*				
	Triglycerides	*Diglycerides*	*Mono-glycerides*	*Glycerol*	*Methyl ether*
0.000	1.000	0.000	0.000	0.000	0.000
50.000	0.784	0.136	0.054	0.025	0.321
100.000	0.629	0.175	0.088	0.108	0.675
150.000	0.519	0.190	0.089	0.201	0.972
200.000	0.440	0.194	0.081	0.284	1.209
250.000	0.383	0.191	0.072	0.354	1.397
300.000	0.340	0.184	0.064	0.413	1.549
350.000	0.307	0.174	0.057	0.462	1.673
400.000	0.282	0.165	0.051	0.503	1.775
450.000	0.261	0.155	0.046	0.538	1.861
500.000	0.244	0.146	0.042	0.568	1.934
550.000	0.230	0.138	0.038	0.594	1.997
600.000	0.218	0.131	0.035	0.617	2.051
800.000	0.182	0.109	0.028	0.681	2.208
1,000.000	0.160	0.095	0.024	0.721	2.306
1,500.000	0.131	0.079	0.019	0.771	2.431
2,000.000	0.119	0.073	0.018	0.790	2.480
2,500.000	0.082	0.065	0.024	0.827	2.597
3,000.000	0.082	0.065	0.024	0.827	2.597
3,600.000	0.082	0.065	0.024	0.827	2.597

methyl ester from 2.56 to 2.58 mol (from 86.7% to 87.4%) with an increase in the reaction temperature from 55°C to 60°C. The time to achieve equilibrium in the reaction (1,200…2,000 seconds) is also significant, as it greatly influences the choice of structural parameters for the reactor for the production of methyl esters. With an increase in temperature to 70°C (Figure 18.2c), there is a decrease in the yield of methyl ether to 2.51 mol (up to 85%).

Figure 18.2a and b shows the effect of temperature on the yield of methyl ether. Obviously, an increase in temperature above 60°C practically does not affect the output of biodiesel.

The experimental studies on the degree of transformation of animal fats into biodiesel in time and their comparison with the data obtained theoretically using the mathematical model allowed confirming that the theoretical results fall into the confidence interval with the probability of 0.95 (Ogorodnikov, Zyska, & Sundetov 2018).

The analysis of the discrete system of equations generated by equations 18.18–18.23 enables to predict (see Figure 18.2a and b) an increase in the yield of methyl ester from 2.56 to 2.58 mol (from 86.7% to 87.4%) with an increase in the reaction temperature from 55°C to 60°C. The time to achieve equilibrium in the reaction (1,200…2,000 seconds) is also significant, as it greatly influences the choice of structural parameters for the reactor for the production of methyl esters. With an increase in temperature to 70°C (Figure 18.2c), there is a decrease in the yield of methyl ether to 2.51 mol (up to 85%).

After finding the most favorable conditions for the process, the most rational ones that will be used for designing the equipment will be selected.

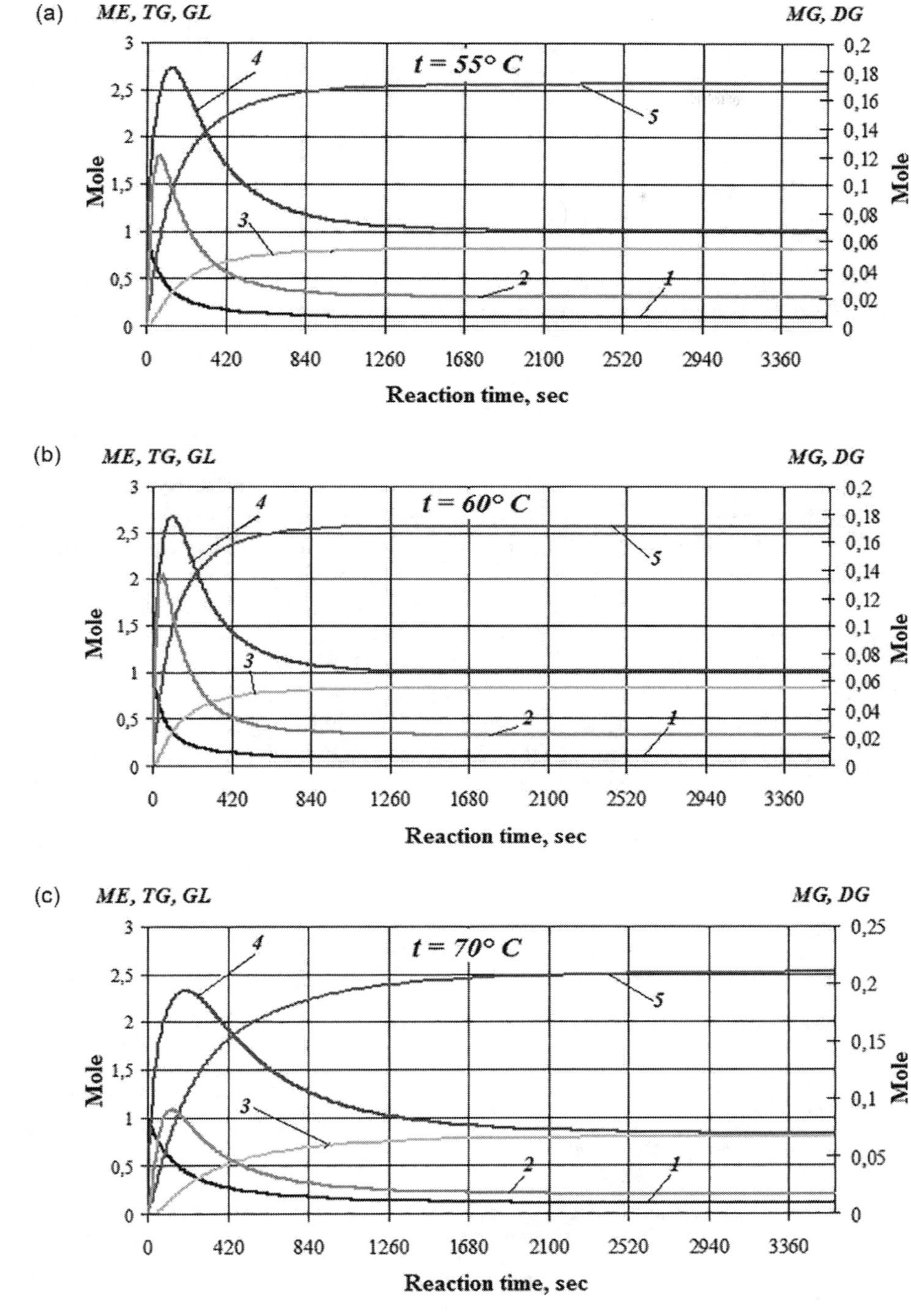

Figure 18.2 (a–c) Effect of temperature on the yield of methyl ether: 1 – triglycerides, 2 – monoglycerides, 3 – glycerol, 4 – diglycerides, and 5 – methyl ether.

Table 18.4 Properties of substances loaded into the reactor

Substance	*Moles*	*Weight, kg*	*Density, kg/l*	*Volume, l*	*Concentration, mol/l*
TG	3,545.2	3,604.2	0.96	3,441.9	0.2045
ME	482,163	1,544.9	0.79	1,955.5	2.7811
B	139,576	5,088.9	1.16	4,886.9	8.05
H_2SO_4	85,089.6	6,646.2	0.88	7,552.5	4.908
Σ	276,428	16,584		17,336	

Table 18.5 Output of the main components of the reaction after 4 hours

Substance	*TG*	*MT*	*B*	*ME*	*DG*	*MG*	*GL*
Initial concentration (mol/l)	0.21	2.78	8.1	0	0	0	0
Final concentration (mol/l)	0.11	2.53	8.1	0.26	0.01	0.01	0.1

After modeling the equations of the molar balance, we turn to their solution using the Matlab software package (the algorithm of the solution presented in Nielsen et al. (2016)). As the initial conditions for the solution of equations, we take the balance of masses. Table 18.4 shows the number of moles, mass, density, and volume in which each of the substances is loaded into the reactor and their initial concentrations. Table 18.5 shows comparison of the main components of the reaction after 4 hours with their initial concentration.

Having analyzed Figure 18.3, it can be concluded that all components of the reaction stabilize after 30 minutes. The transformation of TG reaches 45.52%, indicating that these are not the best conditions for the reactor to work. Such indicators do not meet the technical task, under which the conversion of triglycerides should be equal to 77%.

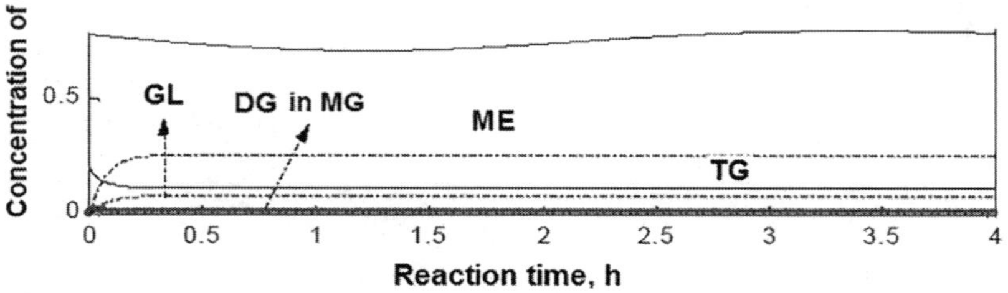

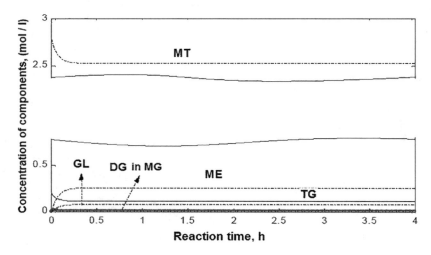

Figure 18.3 Output of the main components of the esterification reaction after 4 hours of process.

Since the conditions we have adopted from the literary sources do not meet our industrial production requirements, we analyze the various concentrations of triglycerides, alcohol, and acids, while analyzing the kinetics of reactions and the variations of transformations for animal fats into biodiesel.

18.4.1 Change in TG Concentration

In order to better analyze the change in the concentration of TG and ME, we neglect the change in the concentration of alcohol, and the reaction time will be reduced to 2 hours: [TG]/2 = 0.10225 (Figures 18.4 and 18.5).

As can be seen from the graphs, the transformation of triglycerides increases with low concentrations and higher.

18.4.2 Change in the concentration of methanol [ME]: [ME]/2 = 1.3905

Having analyzed the graphs in Figures 18.6 and 18.7, it can be concluded that the ether output is proportional to the concentration of methanol, and it increases along with concentrations and vice versa.

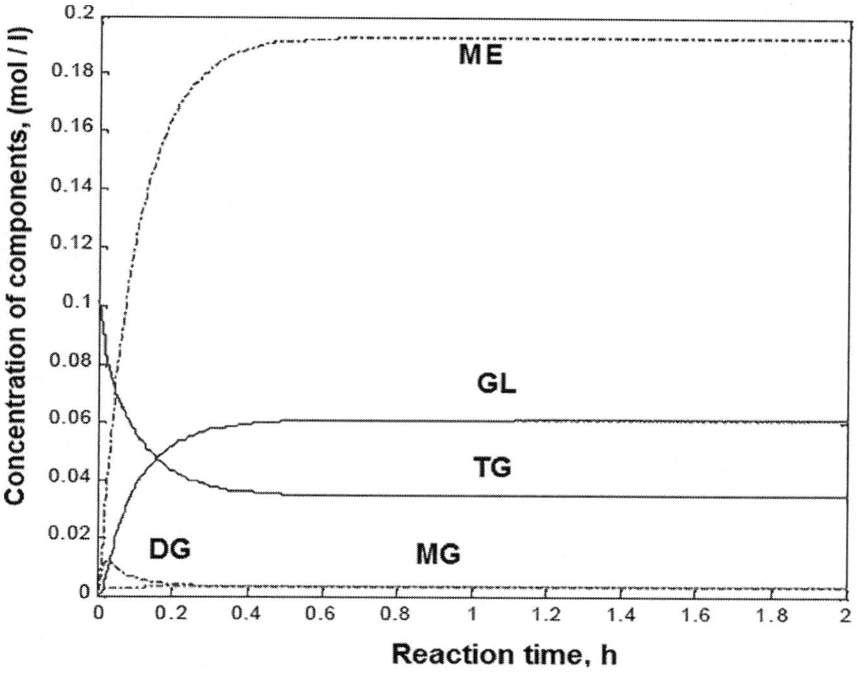

Figure 18.4 Output of the main components of the reaction after 2 hours of the process of changing the concentration of triglycerides: $X = 66.42\%$, [TG] = 0.03433, [ME] = 0.1931, [TG] · 2 = 0.409, where: X – the output of biodiesel.

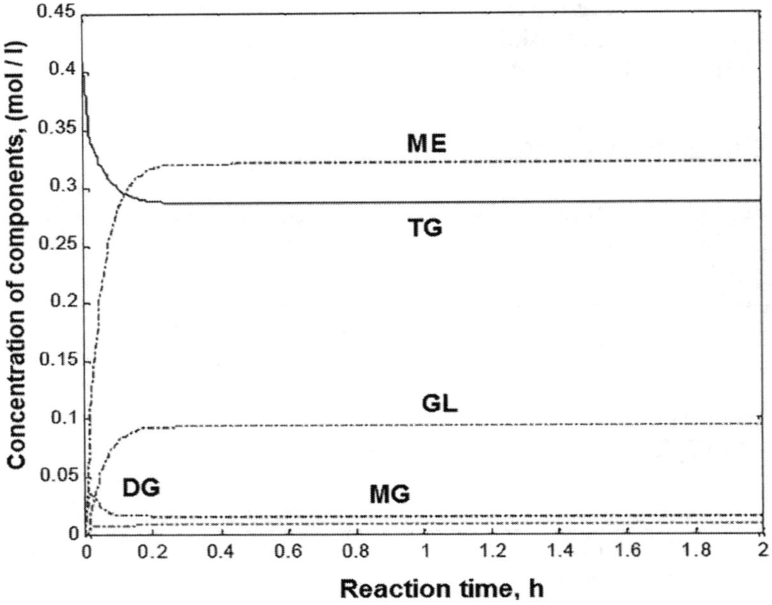

Figure 18.5 Output of the main components of the reaction after 2 hours of the process of changing the concentration of triglycerides: $X = 29.7\%$, $[TG] = 0.2875$, $[ME] = 0.3213$, where: X – the output of biodiesel.

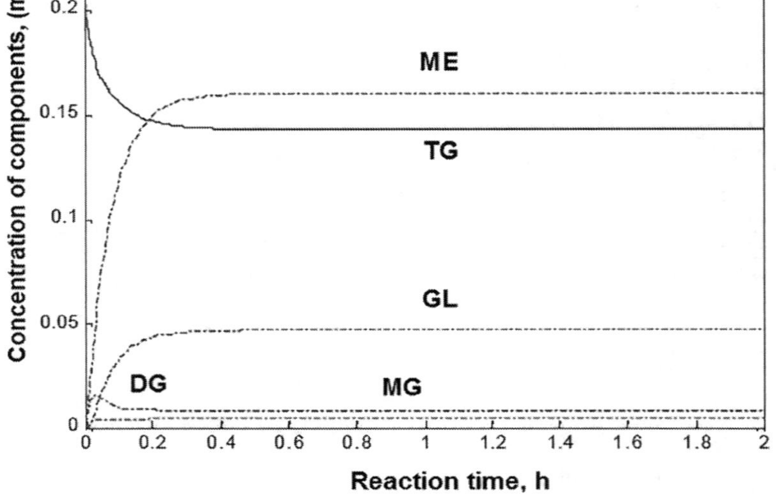

Figure 18.6 Exit of the main components of the reaction after 2 hours of the process of changing the concentration of methanol.

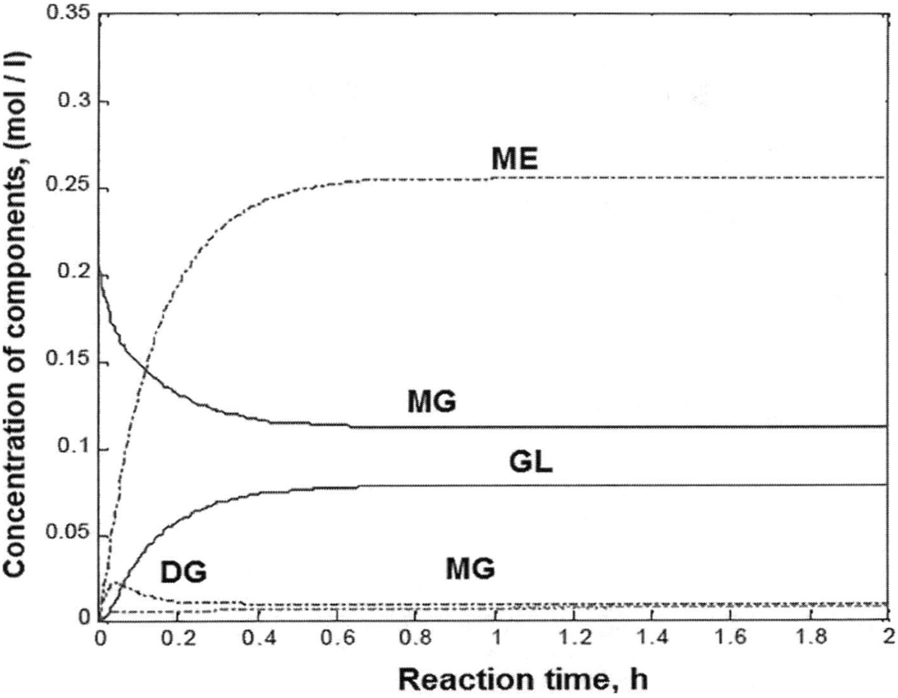

Figure 18.7 Output of the main components of the reaction after 2 hours of the process of changing the concentration of methanol: $X = 66.42\%$, $[TG] = 0.06866$, $[ME] = 0.3862$, where: X – the output of biodiesel.

18.4.3 Change in Catalyst Concentration [H_2SO_4]: [H_2SO_4]/2 = 4.025

If the catalyst concentration changes, the reaction rate does not; in some cases, it is possible to use a higher concentration of the catalyst, but one should not forget that an increase in its concentrations may lead to a decrease in the yield of biodiesel (Figures 18.8 and 18.9). How will the reaction occur when the catalyst concentration is reduced to one-tenth of percentage as shown in Figure 18.10.

If the catalyst concentration changes, the reaction rate does not; in some cases, it is possible to use a higher concentration of the catalyst, but one should not forget that an increase in its concentrations may lead to a decrease in the yield of biodiesel. How will the reaction occur when the catalyst concentration is reduced to one-tenth of percentage as shown in Figure 18.10.

After analyzing the graphs, it can be seen that a significant decrease in the level of components of the reaction leads to an increase in the time they are in the reactor. It can be concluded that the reaction is better at low concentrations of the catalyst and high concentrations of triglycerides and methanol, but the concentration of the catalyst affects the residence time of the mixture in the reactor. In order to have a better idea of the change in the concentrations of the main components of the reaction, the curves of methanol and rice acid concentrations were excluded.

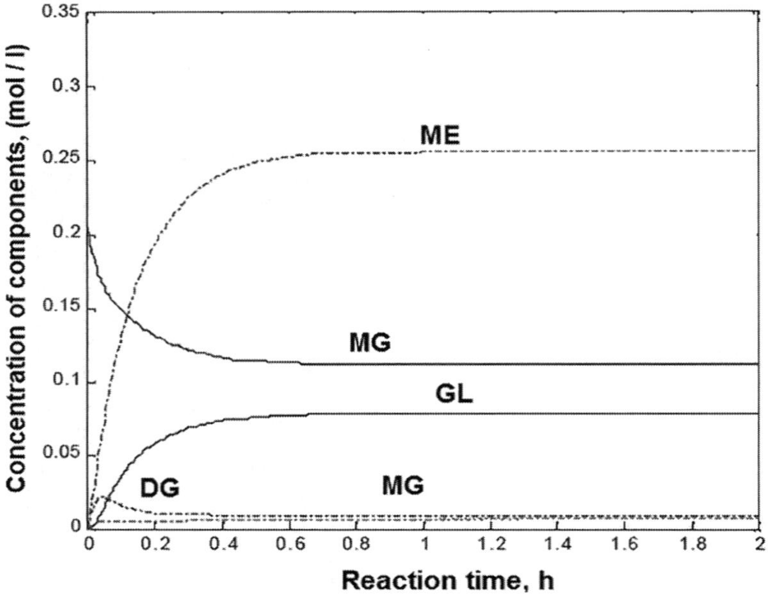

Figure 18.8 Output of the main components of the reaction after 2 hours of process with the change in catalyst concentration: $X = 45.52\%$, [TG] = 0.1114, [ME] = 0.2562, $[H_2SO_4] \cdot 2 = 16.1$, where: X – the output of biodiesel.

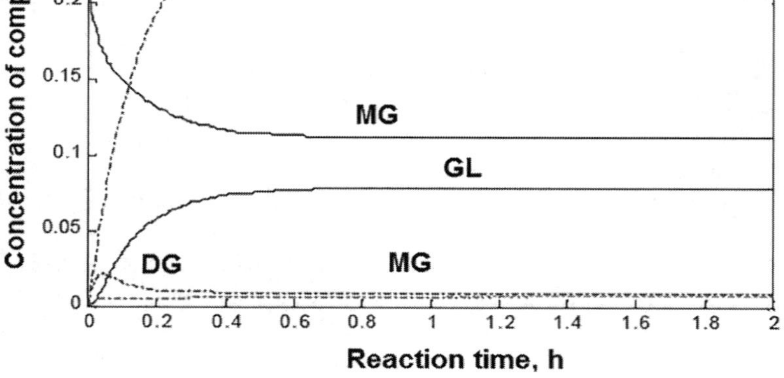

Figure 18.9 Output of the main components of the reaction after 2 hours of the process of changing the catalyst concentration: $X = 45.52\%$, [TG] = 0.1114, [ME] = 0.2562, where X – the output of biodiesel $[H_2SO_4]/10 = 0.805$.

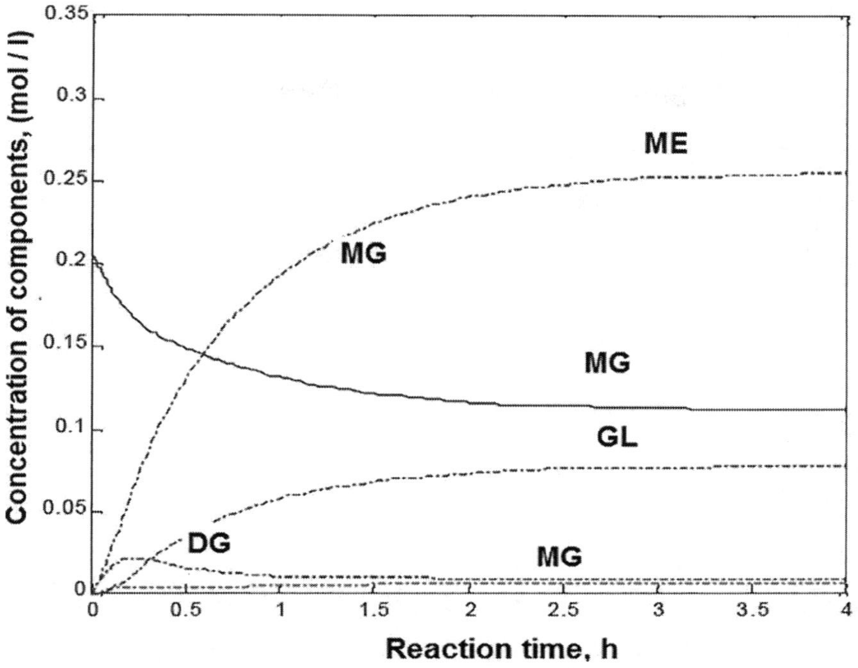

Figure 18.10 Output of the main components of the reaction after 2 hours of process of changing the catalyst concentration: $X = 45.37\%$, [TG] = 0.1117, [ME] = 0.2554, where: X – the output of biodiesel.

Apparently (Figure 18.11), the concentration of methyl ester is rapidly increasing, because it has a higher rate of conversion, the concentration of triglycerides is reduced, but not as fast as diglycerides, the rate of conversion of which increases, and then decreases when converted to monoglycerides and methyl ether. The concentration of monoglycerides increases until the equilibrium of the reaction is achieved. The rate of their conversion is higher than diglycerides conversion rate.

18.5 CONCLUSIONS

1. The developed mathematical model of the reaction of transformation of fats into biodiesel allows using the Microsoft Excel spreadsheet editor with sufficient accuracy to calculate the basic parameters of the reaction of re-esterification.
2. The reaction rate and the completeness of the conversion of components into biodiesel primarily depend on the temperature and composition of the reactant mixture.
3. Increasing the temperature to 65°C accelerates the reaction of transesterification triglycerides, and increases the yield of biofuels, but more than heating reagents to a temperature higher than the boiling point of alcohol (for methanol ~65°C at atmospheric pressure) causes a decrease in the yield of methyl ether.
4. Graphical dependences of the output of the main components of the reaction are constructed when the concentrations of triglycerides, diglycerides, monoglycerides, methanol, catalyst, etc. are changed components of the reaction.

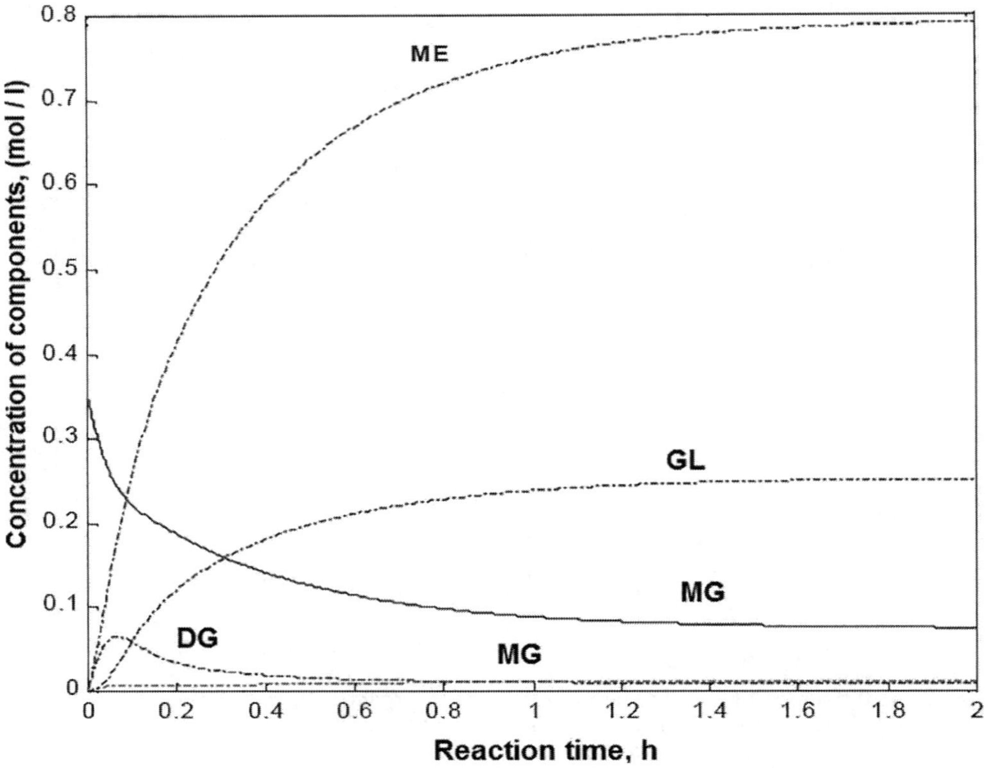

Figure 18.11 Exit of the main components of the reaction after 2 hours of process.

REFERENCES

Adewale, P., Dumont, M.J. & Ngadi, M. 2015. Recent trends of biodiesel production from animal fat wastes and associated production techniques. *Renewable and Sustainable Energy Reviews* 45: 574–588.

Chai, M., Tu, Q., Lu, M. & Yang, Y. 2014. Esterification pretreatment of free fatty acid in biodiesel production, from laboratory to industry. *Fuel Processing Technology* 125: 106–113.

Farooq, M., Ramli, A. & Subbarao, D. 2013. Biodiesel production from waste cooking oil using bifunctional heterogeneous solid catalysts. *Journal of Cleaner Production* 59: 131–140.

Hawrot-Paw, M., Wijatkowski, A. & Mikiciuk, M. 2015. Influence of diesel and biodiesel fuel-contaminated soil on microorganisms, growth and development of plants. *Plant, Soil and Environment* 61(35): 189–194.

Issariyakul, T. & Dalai, A. 2014. Biodiesel from vegetable oils. *Renewable and Sustainable Energy Reviews* 31: 446–471.

Kolyanovska, L.M. et al. 2019. Mathematical modeling of the extraction process of oil-containing raw materials with pulsed intensification of heat of mass transfer. *Proceedings of the SPIE* 11045: 110450X.

Kozlov, L.G. et al. 2019. Experimental research characteristics of counter balance valve for hydraulic drive control system of mobile machine. *Przeglad Elektrotechniczny* 95(4): 104–109.

Mohammadshirazi, A., Akram, A., Rafiee, S. & Kalhor, E.B. 2014. Energy and cost analyses of biodiesel production from waste cooking oil. *Renewable and Sustainable Energy Reviews* 33: 44–49.

Mushtruk, M.M., Sukhenko, Y.G. & Boyko, Y.I. 2017. *Deep Processing of Fats in Byproducts.* Kyiv: Comprint.

Nielsen, P.M., Rancke-Madsen, A., Holm, H.C. & Burton, R. 2016. Production of biodiesel using liquid lipase formulations. *Journal of the American Oil Chemists' Society* 93(7): 905–910.

Ogorodnikov, V.A., Zyska, T. & Sundetov, S. 2018. The physical model of motor vehicle destruction under shock loading for analysis of road traffic accident. *Proceedings of the SPIE* 10808: 108086C.

Polishchuk, L.K. et al. 2018. Study of the dynamic stability of the conveyor belt adaptive drive. *Proceedings of the SPIE* 10808: 1080862.

Polishchuk, L.K. et al. 2019. Study of the dynamic stability of the belt conveyor adaptive drive. *Przeglad Elektrotechniczny* 95(4): 98–103.

Sajith, V., Sobhan, C.B. & Peterson, G.P. 2010. Experimental investigations on the effects of cerium oxide nanoparticle fuel additives on biodiesel. *Advances in Mechanical Engineering* 2: 581–589.

Sukhenko, Y., Sukhenko, V., Mushtruk, M. & Litvinenko, A. 2018. Mathematical model of corrosive-mechanic wear materials in technological medium of food industry. In *Proceedings of the International Conference on Design, Simulation, Manufacturing: The Innovation Exchange, DSMIE-2018*, June 12–15, Sumy, Ukraine, Issue 1: 498–507.

Sukhenko, Y., Sukhenko, V., Mushtruk, M., Vasuliv, V. & Boyko, Y. 2017. Changing the quality of ground meat for sausage products in the process of grinding. *Eastern European Journal of Enterprise Technologies* 4/11(88): 56–63.

Xiong, H., Guo, X. & Xie, W. 2015. Biodiesel remote monitoring system design based on IOT. *Lecture Notes in Computer Science* 8944: 750–756.

Chapter 19

Application of System Analysis for the Investigation of Environmental Friendliness of Urban Transport Systems

Viktoriia O. Khrutba, Vadym I. Zyuzyun, and Oksana V. Spasichenko
National Transport University

Nataliia O. Bilichenko
Vinnytsia National Technical University

Waldemar Wójcik
Lublin University of Technology

Aliya Tergeusizova
Al-Farabi Kazakh National University

Orken Mamyrbaev
Institute of Information and Computational Technologies CS MES RK

CONTENTS

19.1 INTRODUCTION

An efficient city transport system (CTS) is an important factor in ensuring mobile access of the population to workplaces, educational establishments, culture and health care and forms competitiveness of cities. By satisfying the demand of the population in transportation, urban passenger transport affects the level of productivity, domestic services, development of culture and leisure and significantly affects the level of social tension in society. The needs for urban traffic originate in 97% of the population of Ukraine, the annual passenger traffic is more than 70% of the total, and therefore, the improvement of the management of urban passenger

DOI: 10.1201/9781003177593-19

transportation systems becomes crucial for the cities of Ukraine (Shpylyovyi 2010, Zyryanov et al. 2009, Currie et al. 2018).

The papers (Abouhassan 2017, Kharola et al. 2010, Shen et al. 2018) indicate the international experience in the organization of the functioning of the city transport systems, and in the works (Masek et al. 2016, Koetse & Rietveld 2009, Stroe 2014), the influence of the functioning of the city's transport systems on the ecological state of cities and urban environments has been researched (Habr & Veprek 2012, Friman & Fellesson 2009, Marcucci et al. 2011).

The methodological basis for the analysis pertaining to the influence of the functioning of the transport systems of cities on the roadside environment is a systematic approach, which is increasingly used in the research on the impact of urban transport systems and traffic flows, primarily on the environment (Mazzulla & Eboli 2006, Keay & Simmonds 2005, Rakha et al. 2007).

19.2 LITERATURE ANALYSIS AND PROBLEM STATEMENT

System analysis is a method of scientific knowledge, which involves consideration of parts of the object in the inseparable. The methodological basis for system analysis is the notion of "system" – a certain material or ideal object, which is regarded as a complex, holistic entity. The same system can be viewed from different positions, so the system analysis involves the selection of a certain system-generating parameter, which determines the search for a set of elements of the system, a network of relationships and relations between them – its structure (Hidalgo & Huizenga 2013, Martos et al. 2016, Schiavon et al. 2015). Any system is always in a certain environment; because of this system, the analysis should take into account its relationships with the environment. It is mandatory for a system analysis to determine the properties of the whole, while it is emphasized that the essential properties of the system are not determined by the properties of the set of elements, but rather by the properties of its structure, system-forming links of the object (Bertalanffy 1966, Optner 1969, Sadovsky 1974).

The system analysis procedure involves researching the system in order to find the optimal management alternative. It consists of the following stages: the definition of the purpose and objectives of the study and indicators (criteria) of the degree of their achievement; definition of object and subject of the research; purposeful collection and processing of the information on research tasks; definition of the structure of the object, description of its properties, organization and conditions of existence; definition of goals of the life activity of the object; construction of a hypothesis about the mechanism of the object's functioning; object research using models and informal methods, which involves refining the goals and hypotheses about the mechanism of the object's operation, adjustment of models, the definition of a list of possible alternatives to management; and forecasting the consequences of implementing the chosen management alternatives and choosing the most efficient one, that is, making a decision (Stroe 2014, Sharapov et al. 2003).

Researching the system to find the optimal management alternative can be effective with the use of the system analysis procedures. System objects are the input (X_i, number and type of vehicles, road infrastructure, etc.), output (Y_j, economic, social and environmental indicators of transport activity), and the process (organization of the passenger and cargo transportation), which transfers the input parameters into output ($Y_j = f(X_i)$), feedback and restrictions (Z_k, regulatory requirements, financial constraints, technical, organizational, etc.). The system analysis conducted in the study on the city's transport system involves the use of three main approaches that determine the stages of studying systems. They include parametric analysis, morphological analysis, and functional analysis (Mateichik 2014, Zhyuzhyun 2016, Polishchuk et al. 2012).

In accordance with the requirements of the methodology of system analysis in the study of pollution of PS transport streams, three main approaches are used to determine the stages in the study of the system: parametric analysis, morphological analysis, and functional analysis.

Construction of the parametric model of the system allowed determining a group of factors that characterize the state of the system and the conditions of its functioning. The essence of parametric analysis is to substantiate the necessary and sufficient set of factors that allow determining the properties and evaluate the efficiency of the system under investigation (Bryhadyr 2008, Polishchuk et al. 2016, Polishchuk et al. 2018).

Thus, in the study on pollution of the road environment of the TF, the model of interaction between the three subsystems – the traffic flow (TF), the road network (RN) and the road environment (RE), which includes the natural climatic and atmospheric conditions at the point of study, should be considered.

19.3 PURPOSE AND TASKS OF RESEARCH

The essence of parametric analysis is the definition of a group of partial and generalized indicators that characterize all important properties of the "TF – RN – RE" system and the efficiency of its functioning. Parametric synthesis consists in substantiation of the necessary and sufficient number of indicators, which allow estimating the desirable properties of the system (Dmytrychenko et al. 2013, Polishchuk et al. 2019a,b). A generalized parametric analysis of the "TF – RN – RE" system is presented in Table 19.1.

The method of morphological analysis (sometimes called the morphological method) is a combination of the classification method and the generalization method. Its essence lies in the decomposition of the problem into its elements, the search in this scheme of the most perspective on the whole problem element of its solution.

However, a morphological analysis does not involve a simple decomposition, that is, the decomposition of the whole into its constituent parts, but also the selection of elements according to the principles of functional significance and role, that is, the influence of an element or subproblem on the general problem, as well as direct or indirect communication with the external environment (sometimes it is called over-system).

The idea of the method of morphological analysis is to find the largest number, and, to the maximum, and all the possible ways to solve the problem by combining the main structural elements of the systems or signs of solutions, which enables to choose the most expedient of them. Out of all these options, several integer variants are selected as optimal for some target function (Dmytrychenko et al. 2013, Kruze 1981,

Table 19.1 Parametric analysis of the "TF – RN – RE" system

Name of indicator	*Parameter characteristics*
Subsystem "transport flow"	
1.1. Intensity of TP	Main system indicator
1.2. Speed	Main system indicator
1.3. Category T	Main system indicator
1.3.1. Cars (M1)	Partial system indicator
1.3.2. Buses weighing 2.5–5 tons (M2)	Partial system indicator
1.3.3. Buses weighing >5 tons (M3)	Partial system indicator
1.3.4. Trucks with a weight of <3.5 tons (N1)	Partial system indicator
1.3.5. Trucks with a weight of 3.5–12 tons (N2)	Partial system indicator
1.3.6. Lorries weighing >12 tons (N3)	Partial system indicator
1.4. Euro category	Main system indicator
1.4.1. EURO-0	Partial system indicator
1.4.2. EURO-1	Partial system indicator
1.4.3. EURO-2	Partial system indicator
1.4.4. EURO-3	Partial system indicator
1.4.5. EURO-4	Partial system indicator
1.4.6. EURO-5	Partial system indicator
1.5. Type of fuel used by the TP	Main system indicator
1.5.1. Gasoline	Partial system indicator
1.5.2. Diesel	Partial system indicator
1.5.3. Natural gas	Partial system indicator
1.6. Fuel consumption	Main system indicator
1.7. Mode of movement	Main system indicator
1.7.1.An empty passage	Partial system indicator
1.7.2Fixed motion	Partial system indicator
1.7.3.Acceleration mode	Partial system indicator
1.7.4. Braking	Partial system indicator
Subsystem "road network"	
2.1. Width of the travel section	Main system indicator
2.2. Length of the travel part	Main system indicator
2.3. Number of lanes	Main system indicator
2.4. Number of overground pedestrian crossings	Main system indicator
2.5. Number of underground pedestrian crossings	Main system indicator
2.6. Number of traffic lights	Main system indicator
2.7. The value of the coefficient of road resistance	Main system indicator
Subsystem "road environment"	
3.1. Atmospheric conditions	Main system indicator
3.2. Speed and wind direction	Main system indicator
3.3. Density of the strip of green plantations	Main system indicator
3.4. Species composition of plants	Main system indicator
3.5. The width of the strip of green plantations	Main system indicator

Zhelnovach 2012). Structural-morphological description of the "TF – RN – RE" system is shown in Figure 19.1.

Using parametric and morphological analyses, we have constructed a functional system model for identifying environmental hazards that may accompany the functioning of city transport systems (Figure 19.2).

The use of the above-mentioned mechanism will help to more accurately take into account all possible environmental hazards and factors that may accompany the functioning of transport systems of cities and urban areas.

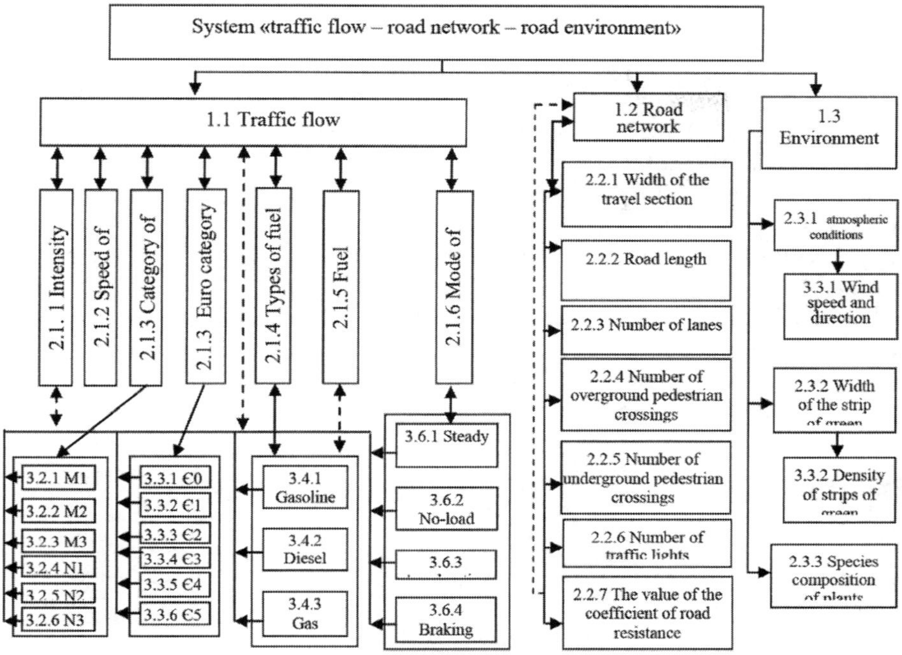

Figure 19.1 Morphological analysis of the "TF – RN – RE" system.

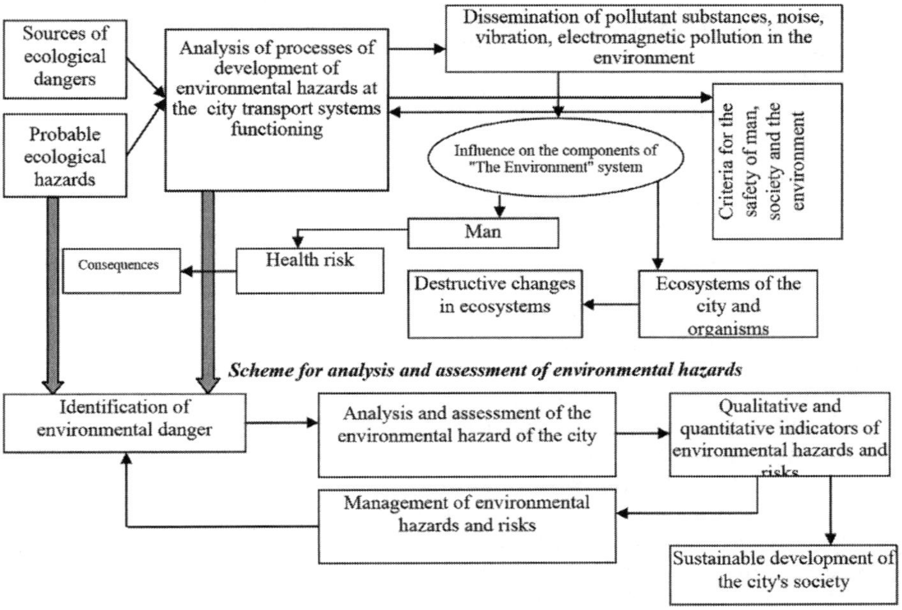

Figure 19.2 System analysis of possible ecological hazards in the process of functioning of transport systems of cities.

19.4 CONCLUSIONS

Thus, on the basis of structural and morphological analyses of the "TF – RN – RE" system for managing the transport system of cities, the "traffic flow", "road network", and "roadside environment" subsystems are allocated. The "traffic flow" subsystem in the second level covers the macro- and micro-parameters the properties of which are determined at the level of system elements. The "traffic flow" subsystem is the most important component, since vehicles are the most significant objects of unfavorable environmental load, namely pollution of atmospheric air. The "road network" subsystem, in conjunction with the "traffic flow" subsystem, constitutes the objects of negative component load. The "roadside environment" subsystem causes the parameters of atmospheric conditions that affect the level of dissemination of pollutants. The morphological model constructed on the basis of parametric analysis allows studying various variants of management of a transport system of cities, and a system model of analysis of possible ecological hazards.

REFERENCES

Abouhassan, M. 2017. Urban transport system analysis. *WIT Transactions of the Built Environment* 176: 57–68. https://www.witpress.com/elibrary/wit-transactions-on-the-built-environment/176/36351

Bertalanffy, L. 1966. *General System Theory*. Moscow: Myr.

Bryhadyr, I.V. 2008. Legal regulation of environmental safety in the field of motor transport: Diss. Cand. Juridicial Sciences: 12.00.06.

Currie, G. et al. 2018. Exploring links between the sustainability performance of urban public transport and land use in international cities. *Journal of Transport and Land Use* 11: 325–342.

Dmytrychenko, M.F. et al. 2013. Mathematical model of software product "Information and analytical system for assessing sideroad environment pollution by transport flows. Certificate of registration of copyright to work No. 51832 of 21.10.2013.

Friman, M., Fellesson, M. 2009. Service supply and customer satisfaction in public transportation: The quality paradox. *Journal of Public Transportation* 12(4): 57–69.

Habr, J., Veprek, J. 2012. *Systemova analyza transporta*. Praha: SNTL.

Hidalgo, D., Huizenga, C. 2013. Implementation of sustainable urban transport in Latin America. *Research in Transportation Economics* 40: 66–77.

Keay, K., Simmonds, I. 2005. The association of rainfall and other weather variables with road traffic volume in Melbourne, Australia. *Accident Analysis & Prevention* 37: 109–124.

Kharola, P.S. et al. 2010. Traffic safety and city public transport system. Case study of Bengaluru, India. *Journal of Public Transportation* 13(4): 63–93.

Koetse, M.J., Rietveld, P. 2009. The impact of climate change and weather on transport: An overview of empirical findings. *Transportation Research Part D: Transport and Environment* 14: 205–221.

Kruze, A.O. 1981. Analysis of the system "City Transport – Urban Environment". Integrated development of road transport in large cities on the example of Moscow. Moscow, 221–222.

Marcucci, E. et al. 2011. *Local Public Transport, Service Quality and Tendering Contracts in Venezia*. Milano: Urban Sustainable Mobilità, 1–14.

Martos, A. et al. 2016. Towards successful environmental performance of sustainable cities: Intervening sectors. A review. *Renewable and Sustainable Energy Reviews* 57: 479–495.

Masek, P. et al. 2016. A harmonized perspective on transportation management in smart cities: The novel IoT-driven environment for road traffic modeling. *Sensors* 16: 1872.

Mateichik, V.P. 2014. Modeling of the system "traffic flow – road" Scientific Notes. Intercollegiate collection (by branches of knowledge "Mechanical engineering and metalworking", "Engineering mechanics", "Metallurgy and materials science"). Lutsk, 2014. Publication No. 46: 371–382.

Mazzulla, G., Eboli, L. 2006. A service quality experimental measure for public transport. *Journal of European Transport* 34: 42–53.

Optner, S.L. 1969. *System Analysis for Solving Business and Industrial Tasks.* Moscow: Soviet Radio.

Polishchuk, L. et al. 2012. The research of the dynamic processes of control system of hydraulic drive of belt conveyors with variable cargo flows. *Eastern-European Journal of Enterprise* Technologies 2(8): 22–29

Polishchuk, L. et al. 2016. Prediction of the propagation of crack-like defects in profile elements of the boom of stack discharge conveyor. *Eastern-European Journal of Enterprise Technologies* 6(1): 44–52.

Polishchuk, L.K. et al. 2018. Study of the dynamic stability of the conveyor belt adaptive drive. *Proc. SPIE* 10808: 1080862. doi:10.1117/12.2501535.

Polishchuk, L.K. et al. 2019a. Study of the dynamic stability of the belt conveyor adaptive drive. *Przegląd Elektrotechniczny* 95(4): 98–103.

Polishchuk, L.K. et al. 2019b. Experimental research characteristics of counter balance valve for hydraulic drive control system of mobile machine. *Przegląd Elektrotechniczny* 95(4): 104–109.

Rakha, H. et al. 2007. *Empirical Studies on Traffic Flow in Inclement Weather; Final Report – Phase I.* Washington, DC: U.S. Department of Transportation.

Sadovsky, V.N. 1974. *Foundations of the General System Theory.* Moscow: Nauka.

Schiavon, M. et al. 2015. Assessing the air quality impact of nitrogen oxide and benzene from road traffic and domestic heating and the associated cancer risk in an urban area of Verona (Italy). *Atmospheric Environment* 120: 234–243.

Sharapov, O.D. et al. 2003. *System Analysis.* Kiev: KNEU.

Shen, L. et al. 2018. Sustainable strategies for transportation development in emerging cities in China: A simulation approach. *Sustainability* 10: 844.

Shpylyovyi, I.F. 2010. *Methodical Foundations of Management of Systems of City Passenger Transportations: Dis. Cand. of Engineering Sciences.* Kyiv: National Transport University.

Stroe, C.C. 2014. Some considerations on the environmental impact of highway traffic. *Revista de Chimie* 65: 152–155.

Zhelnovach, H.M. 2012. Assessment of quality and increase of ecological safety of roadside space: Diss. Candidate of Engineering Sciences: 21.06.01.

Zhyuzhyun, V.I. 2016. System model of environmental risk management in projects. *Bulletin of the National Transport University* 2(35): 84–92.

Zyryanov, V., Keridi, P. Guseynov, R. 2009. Traffic modelling of network level system for large event. In *Proceedings of the 16th World Congress on Intelligent Transport Systems and Services*, Stockholm, Sweden, 21–25 September 2009; Transportation Research Board: Washington, DC, Code 117201.